Space and Society

The Space and Society series explores a broad range of topics in astronomy and the space sciences from the perspectives of the social sciences, humanities, and the arts. As humankind gains an increasingly sophisticated understanding of the structure and evolution of the universe, critical issues arise about the societal implications of this new knowledge. Similarly, as we conduct ever more ambitious missions into space, questions arise about the meaning and significance of our exploration of the solar system and beyond. These and related issues are addressed in books published in this series. Our authors and contributors include scholars from disciplines including but not limited to anthropology, architecture, art, environmental studies, ethics, history, law, literature, philosophy, psychology, religious studies, and sociology. To foster a constructive dialogue between these researchers and the scientists and engineers who seek to understand and explore humankind‘s cosmic context, the Space and Society series publishes work that is relevant to those engaged in astronomy and the space sciences, while also being of interest to scholars from the author‘s primary discipline. For example, a book on the anthropology of space exploration in this series benefits individuals and organizations responsible for space missions, while also providing insights of interest to anthropologists. The monographs and edited volumes in the series are academic works that target interdisciplinary professional or scholarly audiences. Space enthusiasts with basic background knowledge will also find works accessible to them.

The series was founded by Springer together with Founding Editor **Douglas A. Vakoch**.

George D. Kyriakopoulos

Space Resource Utilization and Exploitation: The Suspended Step of International Law

Springer

George D. Kyriakopoulos
School of Law
Kapodistrian University of Athens
Athens, Greece

ISSN 2199-3882 ISSN 2199-3890 (electronic)
Space and Society
ISBN 978-3-031-96335-3 ISBN 978-3-031-96336-0 (eBook)
https://doi.org/10.1007/978-3-031-96336-0

This Springer imprint is published by the registered company Springer Nature Switzerland AG
The registered company address is: Gewerbestrasse 11, 6330 Cham, Switzerland

To my father, who will always live in me.

Foreword

Public international law has as its *raison d'etre* the promotion of international peace and security. Although challenges to the global public order exist in multiple contexts, the potential for disruptions to international relations is especially present in endeavors which lack a general consensus on the applicable legal rules. Such is the situation facing the nascent but burgeoning commercial space industry. Private enterprise is poised to exploit the capabilities pioneered by government space programs and literally take commerce to new and unimaginable heights on the Moon, asteroids, and beyond. It is widely believed that these celestial bodies could contain vast riches of resources that can be used for the benefit of mankind, yet legal rules and regulations have not kept pace with the technological capabilities of the private sector to discover and exploit these resources.

To date, the governance of private enterprise in space has consisted of principles set forth in the Outer Space Treaty of 1967, with some influence by the Moon Agreement of 1979, and a gallimaufry of soft law statements, national laws, and international studies and accords. Prof. George D. Kyriakopoulos has taken on the ambitious task of trying to sort through this maze with a fresh perspective and succeeded in producing a concise and elucidating exposition of the dilemma facing international space law, with a pragmatic solution, presented in a style that is thorough while not excessively technical.

Prof. Kyriakopoulos has dismantled the trap that so many other commentators have fallen into that frame the discussion in terms of "property rights" in space resources. All too often proposals are advanced that engage in convoluted logic and elaborate rationalizations to justify a pre-ordained position that sanctions various traditional forms of property rights sought to be claimed by the private sector in space resources notwithstanding the non-appropriation provision in Article II of the Outer Space Treaty. Prof. Kyriakopoulos charts a different course.

In this book, Prof. Kyriakopoulos reviews the inadequacy of the domestic legislation of states that purport to grant or recognize private sector property rights in space resources *vis-a-vis* the non-appropriation principle as expressed in both the Outer Space Treaty and the Moon Agreement. He also provides a detailed and insightful discussion of the diplomatic activities of the member states in the United

Nations Committee on the Peaceful Uses of Outer Space leading to the establishment of the current Working Group on Space Resources, tempered by the fact that there has not been a new treaty devoted to space since the Moon Agreement in 1979. Prof. Kyriakopoulos critically examines the Building Blocks of the Hague International Space Resources Governance Working Group, noting that the fundamental principles set forth therein regarding priority rights, resource rights, and safety zones do not reflect *de lege lata* but are *de lege ferenda* and will require substantial revision of current space law if they are to be implemented. The Artemis Accords also are analyzed, with their implicit emphasis on promoting a state practice recognizing property rights in space resources as a predicate to claiming legal acceptance of this activity through having acquired customary force.

After identifying the inadequacies of unilateral state legislation, the protracted difficulty COPUOS experienced in obtaining consensus to establish the Working Group on Space Resources and its inability to produce a new treaty in more than 40 years, and the inherent weaknesses of the Hague Building Blocks and the Artemis Accords, Prof. Kyriakopoulos arrives at a pragmatic and thought-provoking solution that invokes the jurisdiction of the International Court of Justice.

Prof. Kyriakopoulos is well qualified to take on this task, as an internationally recognized and respected expert on space law. He has been an inspiration to students around the world, including as the coach of numerous moot court teams from the School of Law of the National and Kapodistrian University of Athens and the author of very successful cases for the Manfred Lachs Space Law Moot Court Competition, where I have had the pleasure of working with him as a Co-Chair of the International Institute of Space Law Moot Court Committee. In that role, I particularly appreciated his most recent subtle and clever incorporation of the names of some of his favorite students into the moot court fact situation.

Prof. Kyriakopoulos has included the full texts of the Outer Space Treaty and other pertinent instruments and materials as a useful reference. His book flows like many conversations he and I have had, whether discussing the intricacies and nuances of space law or the sublime elegance of the music of the Allman Brothers Band. This book should inspire many conversations in the space law community and beyond.

Law Offices of Sterns and Tennen
Phoenix, AZ, USA

Leslie I. Tennen

Acknowledgments

I am grateful to my research assistants:

Ms. Niki Giannakou, Attorney at Law, Athens Bar Association, LL.M. National Kapodistrian University of Athens,

Mag. iur. Arthur Gehr, LL.M. Leiden University,

Ms. Nariat Pashaeva, LL.M. University of Oslo,

Ms. Foteini Daliani, LL.M. National and Kapodistrian University of Athens, LL.M. UCLA,

Mr. Alexandros Vourdas, Attorney at Law, Athens Bar Association, LL.M. National Kapodistrian University of Athens,

who supported this project by undertaking focused research on assigned topics under my direction.

Contents

Abbreviations

A.A.S.L.	American Association of School Librarians
A.S.I.L.	American Society of International Law
Adv. S. Res.	Advances in Space Research
art.	Article
CEO	Chief executive officer
COPUOS	Committee on the Peaceful Uses of Outer Space
COSPAR	Committee on Space Research
COVID	SARS-CoV-2 (2019-nCoV) coronavirus
Doc.	Document
Dr.	Doctor
ECOSOC	Economic and Social Council
Ed.	Editor
Eds.	Editors
ESA	European Space Agency
Et seq.	et sequentia
GA	General assembly
GEO	Geostationary orbit
GRULAC	Group of Latin America and the Caribbean
GSO	Geosynchronous orbit
H.R.	Human resources
HEO	Highly elliptical orbit
HG	Hague International Space Resources Governance Working Group
i.e.	*id est*
IAC	International Astronautical Congress
ICAO	International Civil Aviation Organization
ICJ	International Court of Justice
ID	Identity
IISL	International Institute of Space Law
ISRU	*In situ* resource utilization
ITLOS	International Tribunal for the Law of the Sea
ITU	International Telecommunication Union

J. Space L.	Journal of Space Law
J.I.L.	Journal of International Law
J.S.L.	Journal of Space Law
L.J.	Law Journal
LSC	Legal Subcommitte (COPUOS)
MEO	Medium Earth Orbit
Mr.	Mister
NASA	National Aeronautics and Space Administration
NEO(s)	Near-Earth Object(s)
No.	Number
Op. cit.	Opus citatum
OS	Outer space
OST	Outer Space Treaty
p.	Page
par.	Paragraph
PCIJ	Permanent Court of International Justice
Res.	Resolution
SC	Security Council
STSC	Scientific and Technical Subcommittee (COPUOS)
U.A.E.	United Arab Emirates
U.S.	United States
UK	United Kingdom
UN	United Nations
UNCLOS	United Nations Convention on the Law of the Sea
UNCOPUOS	United Nations Committee on the Peaceful Uses of Outer Space
UNGA	United Nations General Assembly
USA	United States of America
USSR	Union of Soviet Socialist Republics
v., vs.	Versus
VCLT	Vienna Convention on the Law of Treaties
vol.	Volume
WG	Working group

Chapter 1
Introduction

In recent years, the development of activities related to the exploration, exploitation, and use of outer space's natural resources has been the subject of continuous and intense debate. While this constitutes a space activity oriented toward the future—requiring not only technological advancement but also the mobilization of substantial financial resources—these prerequisites alone are insufficient. A further indispensable condition is the establishment of a legal framework capable of effectively addressing and regulating the complex issues that inevitably arise.

In this evolving landscape, new space actors—primarily from the private sector—are stepping into domains that, until recently, were regarded as largely "exotic." This shift presents significant legal challenges, as the existing body of international space law, shaped during the 1950s to 1970s, was originally designed to govern activities conducted almost exclusively by States.

It is generally accepted that the term "space resource utilization" includes: (a) activities in Earth orbits; (b) *in situ* resource utilization (ISRU); and (c) the extraction and appropriation of natural space resources for commercial purposes (the so-called space mining). The definition of exactly which activities need to be regulated, and which space resources will be subject to regulation, is currently being considered by the Working Group on Legal Aspects of Space Resource Activities of the Legal Subcommittee (LSC) of the United Nations Committee on the Peaceful Uses of Outer Space (COPUOS).[1] Said working group was created in the context of the agenda item of the LSC "General exchange of views on potential legal models for activities in the exploration, exploitation and utilization of space resources."[2]

[1] Mainly since the 62nd Session of the UNCOPUOS Legal Subcommittee (20–31 March 2023).

[2] For the definitions of "resources," "exploration," and "exploitation" as they have evolved in the context of activities on the international seabed, see the relevant developments in Chap. 6.

G. D. Kyriakopoulos, *Space Resource Utilization and Exploitation: The Suspended Step of International Law*, Space and Society,
https://doi.org/10.1007/978-3-031-96336-0_1

In relation to these activities, the following should, in the first place, be noted: It is accepted, since the inception of human activity in outer space, that placing a space object in orbit is a lawful activity under international law and space law, compatible with the fundamental principles of space law, such as the principle of freedom of use and exploration of space and the principle of non-appropriation. More than 70 States currently have at least one satellite in orbit (with a steadily increasing trend),[3] and the relevant technology—or the associated provision of services by advanced space-faring States or international organizations—is fully accessible to most States.

In situ resource utilization (ISRU) is the collection, processing, storing, and use of materials encountered during human or robotic space exploration that replace materials that would otherwise be brought from Earth to accomplish a mission critical need at reduced overall cost and risk.[4] ISRU, provided that it does not conflict with Article IX of the 1967 Outer Space Treaty (OST) (which will be considered *ad hoc*), is an activity which falls within the limits of Articles I(1)(b), II and III OST, as it can be justified in the light of the freedom of use, exploration and scientific investigation of outer space, and, in principle, does not constitute an acquisition of property. In any case, a future *ad hoc* regulation of the ISRU could mitigate any potential tensions, should such use be made by numerous users in a common area on a celestial body.

A relevant provision concerning ISRU is included in U.S. Executive Order 13914 of April 6, 2020 on "Encouraging International Support for the Recovery and Use of Space Resources," which, *inter alia*, states that: "Successful long-term exploration and scientific discovery of the Moon, Mars, and other celestial bodies will require partnership with commercial entities to recover and use resources, including water and certain minerals, in outer space."[5]

[3] According to pixalytics.com (which uses data from the Union of Concerned Scientists (UCS)), in the beginning of 2023, there were 6718 operational satellites into orbit, which corresponds to 67.28% of the total orbiting objects. the same source counts 3266 useless objects which can be considered as space debris. The United States controls most of these satellites (4511), followed by China (586), the United Kingdom (561), Russia (177) and India (62). Source: https://www.pixalytics.com/satellites-orbiting-earth-2023/ (last visit on 23.1.2024).

[4] Sackstender K.R. & Sanders G.B., "In-Situ Resource Utilization for Lunar and Mars Exploration", 45th AIAA Aerospace Sciences Meeting and Exhibit, 08 January 2007–11 January 2007, Reno, Nevada, in: https://arc.aiaa.org/doi/pdf/10.2514/6.2007-345 (last accessed: 23.01.2024).

[5] https://www.federalregister.gov/documents/2020/04/10/2020-07800/encouraging-international-support-for-the-recovery-and-use-of-space-resources (last accessed: 23.01.2024).

1.1 The Prospect of Mining and Commercial Exploitation of the Natural Space Resources of the Celestial Bodies

According to NASA, there are about 100,000 Near-Earth Objects (NEOs)[6] and a large number of them potentially contain water and important minerals, such as nickel, cobalt, and iron. However, is "space mining" of these resources currently feasible based on existing technological capabilities?

On 6 August 2014, the European Space Agency's (ESA) Rosetta spacecraft arrived at Comet 67P/Churyumov-Gerasimenko and the Philae space shuttle landed on the comet on 12 November 2014. On 8 September 2016, NASA launched the OSIRIS-REx spacecraft with the mission to approach and land on the asteroid Bennu. The spacecraft returned to Earth on 24 September 2023 and dropped off material from Bennu into the Utah desert, before embarking on a new mission to explore the asteroid Apophis (OSIRIS-APEX).[7] These missions, although experimental, clearly pave the way to the commercial exploitation of the resources of celestial bodies (comets, asteroids, and planets) in the near future.

During the Cold War, the competition between the United States and the Soviet Union to "conquer" outer space resulted in a set of fundamental principles, first enshrined in United Nations General Assembly Resolution 1962 (XVIII) of 13 December 1963 and then codified in the 1967 Outer Space Treaty (OST). In this context, the need to seek a balance—initially between the—then—two superpowers, which were the only ones operating in outer space—on possible claims to the existing "wealth" in space led to the adoption of the principle of non-appropriation (Article II OST). At that time, the existing technological capabilities did not allow for the exploitation of space mineral resources. Today, however, the circumstances have changed, and it is highly likely that in the near future we will witness growing competition among States for access to and control over the mineral resources of celestial bodies. This prospect clearly underscores the need for the establishment of a global regulatory regime governing such resources, as unilateral actions are likely to generate significant tensions among the States involved. Any such regime must remain consistent with the foundational principles of international space law, which, to date, have not been formally contested by States.

---

[6] See NASA, Near-Earth Object Survey and Deflection Analysis of Alternatives, Report to Congress, March 2007, in: https://www.nasa.gov/wp-content/uploads/2015/01/171331main_neo_report_march07.pdf (last accessed: 28.01.2024). As designated by NASA, "Near-Earth Objects (NEOs) are comets and asteroids that have been nudged by the gravitational attraction of nearby planets into orbits that allow them to enter the Earth's neighborhood. Composed mostly of water ice with embedded dust particles, comets originally formed in the cold outer planetary system while most of the rocky asteroids formed in the warmer inner solar system between the orbits of Mars and Jupiter. The scientific interest in comets and asteroids is due largely to their status as the relatively unchanged remnant debris from the solar system formation process some 4.6 billion years ago"—https://cneos.jpl.nasa.gov/about/basics.html (last accessed: 28.01.2024).

[7] See https://www.nasa.gov/osiris-rex (last accessed: 28.01.2024).

This is the central question addressed in this book: how to agree on an effective regime for the management of space resources that will be universally accepted by the international community. Moreover, what challenges would such a regime pose to existing international space law.

As will be shown below, the *Magna Carta* of space law, the Outer Space Treaty of 1967, does not contain any specific provisions on this issue, since the priorities of the two superpowers, the United States and the Soviet Union, who were the main space players of the time, were different. A few years later, the Moon Agreement of 1979 would consider the Moon and its natural resources as "the common heritage of mankind" and would, among other things, provide for an international regime "to govern the exploitation of the natural resources of the Moon as such exploitation is about to become feasible." However, although this international treaty has entered into force, it has so far been ratified by only 17 States.[8] Its provisions are therefore far from describing a generally accepted international legal framework on the exploitation of space resources.

Conversely, we are now living in the era of "NewSpace"[9] activities: this term describes the increasing involvement of the private sector in the space domain, mainly through commercial applications. The provision of services *via* satellites, space tourism, and the transport of cargo and people to the International Space Station are—so far—the main manifestations of a trend toward the commercialization of space, in which private operators are increasingly involved. Clearly, in the context of NewSpace, the issue of space resources is inherently important to the emerging "space economy"[10] and therefore of great interest to stakeholders in the new space era.

The absence of a solid and convincing international legal framework has led some countries, such as the United States or Luxembourg, to take unilateral legislative initiatives in order to create an environment of legal certainty and thus give the necessary impetus to the funding of the research and technology required to make the exploitation of the resources of the celestial bodies technically feasible in the

[8] On 1 January 2023, 18 States had ratified the Moon Agreement. However, on 5 January 2023 Saudi Arabia notified its withdrawal from the treaty, a withdrawal which took effect on 5 January 2024—see https://www.unoosa.org/oosa/en/ourwork/spacelaw/treaties/status/index.html (last accessed: 28.01.2024).

[9] See: https://www.esa.int/About_Us/Business_with_ESA/Global_Space_Economic_Forum/New_Space_and_the_role_of_public_support; https://professionalprograms.mit.edu/blog/technology/what-is-new-space-economy/ (last accessed: 28.01.2024).

[10] The main features of the modern space economy are increasing private investment, the proliferation of start-ups, and the intense activity of large private companies (such as Space X). As one of the four pillars defined in the Dubai Declaration (adopted at the first High Level Forum: Space as a driver for socio-economic sustainable development, Dubai, United Arab Emirates, 24 November 2016), the space economy has a special place in the context of the Space2030 agenda. See: "What is the New Space Economy?" https://professionalprograms.mit.edu/blog/technology/what-is-new-space-economy/; "What is the Space Economy?" https://space-economy.esa.int/article/33/what-is-the-space-economy; OECD Handbook on Measuring the Space Economy, 2nd Edition, 2022, https://doi.org/10.1787/8bfef437-en.

short term. However, as will be shown below, the legitimacy of such initiatives, from the point of view of international law, is highly questionable and remains in the minority on the international scene, at least for the time being.

1.2 In Search of a Legal Framework: International Law in Limbo

The combination of the above conditions has led to an impasse on the issue of regulating the management of space resources, further complicated by the fact that the international space community appears entirely unwilling to move toward the adoption of a new binding international instrument after the 1979 Moon Agreement. In this context, the current state of international law could be likened to the old, iconic film by Director Theo Angelopoulos, *The Suspended Step of the Stork* (1991), where the protagonists stand on a dubious step, at the borderline, facing an uncertain path:

> The titular step comes from two bookend scenes. In the first, a Greek colonel walks along a bridge linking the two countries and stops right at the frontier line. In the distance we see rifle-toting Albanian guards. The colonel holds his right foot in the air hovering above the line. He says, "Greece ends at the blue line. *If I take one more step I will be somewhere else, or die.*" At the end of the film the Greek television reporter repeats this gesture (as does the symbolic journey, which both the mythical Ulysses figure played by Marcello Mastroianni and the journalist take). This sense of repetition, of history repeating itself and the journey commencing once again, is, in one sense, also the film's source of hope (emphasis added).[11]

Accordingly, international law is currently at crossroads regarding its future direction on the issue of space resources. The diversity of views and positions among States, the rigidity of existing space law, and the reluctance of States to undertake new conventional commitments at the international level all contribute to a state of legal stagnation and inertia.

However, Article III of the Outer Space Treaty of 1967 states that:

> States Parties to the Treaty shall carry on activities in the exploration and use of outer space, including the Moon and other celestial bodies, *in accordance with international law*, including the Charter of the United Nations, in the interest of maintaining international peace and security and promoting international cooperation and understanding (emphasis added).

Consequently, the primacy of international law must be restored, and international law must find a way forward. It must take a next, decisive step in the governance of the use and exploitation of space resources. This book, after first navigating through the applicable law and highlighting its structural and substantial lack of adequacy (Chap. 2), further documents whether the current impasse will ultimately prove irreparable, despite the systematic pursuit of suitable interpretative approaches by

[11] Donato Totaro, "The Suspended Step of the Stork: Theo Angelopoulos among the Greats", Off Screen, Volume 20, Issue 2, February 2016, https://offscreen.com/view/the-suspended-step-of-the--stork_theo-angelopoulos (last visit on 28.1.2024).

space law scholars (Chap. 3) and the active engagement of States—both through their participation in relevant international *fora* (Chap. 4) and their promotion of international cooperation *via* agreements, though not on a global scale (Chap. 5). Finally, this book explores the conditions, ways, and means under which the desired, firm step of international law can take place (Chap. 6).

Chapter 2
Realizing the Need to Take a Step: Applicable International Law Is Not Enough

As noted in the Introduction, significant legal deadlocks persist in relation to the exploration and use of space resources. These impasses stem primarily from the absence of explicit provisions in the existing *corpus* of international space law concerning space resources. At the same time, national legislative efforts remain inadequate, as they often diverge from or are incompatible with the fundamental principles of international space law.

2.1 Perplexity of International Instruments

2.1.1 The 1967 Outer Space Treaty

In connection with the law governing space resource activities, the Outer Space Treaty of 1967,[1] which is largely accepted by the international community as the *magna carta* of space law, has raised heated debates concerning the content of rights and obligations envisaged in its Articles I and II. These two Articles of the OST are fundamental for the legal regime governing space activities, given that they establish the rights and freedoms of States in outer space, while at the same time providing for limitations to these freedoms.

First, as already stressed, according to Article III OST, activities for the exploration and use of outer space, including the Moon and other celestial bodies, must be conducted in conformity with International Law:

[1] Treaty on Principles Governing the Activities of States in the Exploration and Use of Outer Space, including the Moon and Other Celestial Bodies, entered into force Oct. 10, 1967, 18 U.S.T. 2410, 610 U.N.T.S. 205, [hereinafter OST].

G. D. Kyriakopoulos, *Space Resource Utilization and Exploitation: The Suspended Step of International Law*, Space and Society,
https://doi.org/10.1007/978-3-031-96336-0_2

> Article III
> States Parties to the Treaty shall carry on activities in the exploration and use of outer space, including the moon and other celestial bodies, in accordance with international law, including the Charter of the United Nations, in the interest of maintaining international peace and security and promoting international co-operation and understanding.

In this context, the main applicable provisions relating to the exploration and use of outer space, are Articles I and II OST. Article I provides for freedom of exploration and use of outer space, including the Moon and other celestial bodies, and freedom of access to all regions of celestial bodies:

> Article I
> The exploration and use of outer space, including the moon and other celestial bodies, shall be carried out for the benefit and in the interests of all countries, irrespective of their degree of economic or scientific development, and shall be the province of all mankind.
> Outer space, including the moon and other celestial bodies, shall be free for exploration and use by all States without discrimination of any kind, on a basis of equality and in accordance with international law, and there shall be free access to all areas of celestial bodies.
> There shall be freedom of scientific investigation in outer space, including the moon and other celestial bodies, and States shall facilitate and encourage international co-operation in such investigation.

… whereas Article II introduces a restriction on the exercise of the freedoms guaranteed in Article I, by establishing the principle of non-appropriation:

> Article II
> Outer space, including the moon and other celestial bodies, is not subject to national appropriation by claim of sovereignty, by means of use or occupation, or by any other means.

It follows that the combination of Articles I and II OST renders outer space and its resources *res communes omnium*, hence subject to access and use by anyone, in exclusion of exercise of sovereignty or property rights over them.

On this note, the interpretation of Article II OST has caused great controversy throughout the international community with respect to potential mining activities on the moon and other celestial bodies for commercial purposes. The conflicting views on this issue are evident both at the level of legal theory and at the level of state practice.

2.1.1.1 The Non-appropriation Principle *vs.* Property Rights in Outer Space

As already mentioned, Article I para. 2 of the OST establishes the principle of freedom of exploration and use of the outer space. This "freedom of exploration and use" is not limitless: It must be interpreted in the light of art. II of the same treaty, which sets out the principle "of non-appropriation". Same principle is contained in art. 11 para. 2 of the 1979 Moon Agreement:[2]

[2] Agreement Governing the Activities of States on the Moon and Other Celestial Bodies, adopted on 5 December 1979, opened for signature on 18 December 1979, entered into force on 11 July 1984. 1363/U.N.T.S./3 (hereinafter "The Moon Agreement" or "MOON"). It is important that, in accordance with art. 1 of the Moon Agreement, references to the moon in the treaty shall be understood as applicable to all celestial bodies within the solar system, other than the Earth.

> The Moon is not subject to national appropriation by any claim of sovereignty, by means of use or occupation, or by any other means.

The core element in art. II of the OST is "national appropriation". Said expression existed in UNGA Resolution 1721(XVI) of 20 December 1961 as well as in UNGA Resolution 1962(XVIII) of 13 December 1963 (principle No 3). According to the opinions expressed during the *travaux préparatoires* of the Outer Space Treaty—and those of distinguished scholars—,[3] the expression "national appropriation" prohibits both the exercise of sovereign rights (by States) and private appropriation (by non-governmental entities). This conclusion is further strengthened by the clear wording of art. 11 para. 3 of the Moon Agreement, according to which

> Neither the surface nor the subsurface of the Moon, nor any part thereof or natural resources in place, *shall become property of any State, international intergovernmental or non-governmental organization, national organization or non-governmental entity or of any natural person*[4] (emphasis added).

Accordingly, the establishment of property rights in outer space—by States or private actors—remains difficult to reconcile with the non-appropriation principle. Of course, in accordance with Article VIII para. 2 of the OST, "ownership of objects launched into outer space, including objects landed or constructed on a celestial body, and of their component parts, is not affected by their presence in outer space or on a celestial body or by their return to the Earth". However, this article is inapplicable against resources mined on a celestial body, since they were previously covered by the non-appropriation principle and, obviously, ownership thereon should be inconceivable.[5]

Finally, Article IX OST calls for an increased multilateralism in space affairs, stressing the importance of international cooperation and undertaking in outer space "with due regard to the corresponding interests of other States". Along the same lines, Article IX, *in fine*, requires consultations in the event that a planned activity in outer space may cause harmful interference with the activities of other States:

---

[3] S. Freeland & R. Jakhu, "Article II", in S. Hobe, B. Schmidt-Tedd, K.-U. Schrogl & G. Meishan Goh (eds.), Cologne Commentary on Space Law, Vol. 1 Outer Space Treaty, Carl Heymanns Verlag, 2009, p. 50; S. Hobe, "Adequacy of the Current Legal and Regulatory Framework Relating to the Extraction and Appropriation of Natural Resources", in Institute of Air and Space Law, McGill University, Policy and Law Relating to Outer Space Resources: Examples of the Moon, Mars, and other Celestial Bodies, Workshop Proceedings, 28–30 June 2006, p. 206; S. Hobe, "The Legal Framework for a Lunar Base Lex Data and Lex Ferenda", in G. Lafferranderie & D. Crowther (eds.), Outlook on Space Law over the Next 30 Years: Essays Published for the 30th Anniversary of the Space Treaty, Springer, 1997, p. 138–139.

[4] Emphasis added.

[5] The above remarks on the application of Article II OST, already expressed by the author ("Positive Space Law and Privatization of Outer Space: Fundamental Antinomies", in: G.D. Kyriakopoulos & M. Manoli (eds), The Space Treaties at Crossroads: Considerations de Lege Ferenda, 2019, Springer, Cham, p. 1–13.) are reproduced here, since, in his view, they retain all their relevance at the present stage of development of the issue of exploration, utilization and exploitation of space resources.

Article IX

In the exploration and use of outer space, including the Moon and other celestial bodies, States Parties to the Treaty shall be guided by the principle of co-operation and mutual assistance and shall conduct all their activities in outer space, including the Moon and other celestial bodies, with due regard to the corresponding interests of all other States Parties to the Treaty… If a State Party to the Treaty has reason to believe that an activity or experiment planned by it or its nationals in outer space, including the Moon and other celestial bodies, would cause potentially harmful interference with activities of other States Parties in the peaceful exploration and use of outer space, including the Moon and other celestial bodies, it shall undertake appropriate international consultations before proceeding with any such activity or experiment. A State Party to the Treaty which has reason to believe that an activity or experiment planned by another State Party in outer space, including the Moon and other celestial bodies, would cause potentially harmful interference with activities in the peaceful exploration and use of outer space, including the Moon and other celestial bodies, may request consultation concerning the activity or experiment.

2.1.1.2 Interests of Humankind *vs.* Interests of States and individuals

Humankind (Mankind in treaty wording) holds a prominent place in the outer space treaties: According to Article I para. 1 of the Outer Space Treaty, "The exploration and use of outer space, including the Moon and other celestial bodies, shall be carried out for the benefit and in the interests of all countries, irrespective of their degree of economic or scientific development, and shall be the province of all mankind". An analogous provision can be found in Article 4 para. 1 of the Moon Agreement. Besides, Article V of the OST considers astronauts as "envoys of mankind in outer space", while Article 11 para. 1 of the MOON stipulates that "The Moon and its natural resources are the common heritage of mankind".

Based on these provisions, humankind *per se* may validly be regarded as a distinct subject of international space law.[6] If accepted, this assumption would necessarily imply that every State has the right to require that any activity in outer space be conducted in the interest of all States, or more specifically, in the interest of humankind.[7]

Nevertheless, some scholars consider that a distinction must be made between the two concepts: "Province of all mankind", "Common heritage of mankind". It is obvious that the supporters of private profit-making activities in space clearly adopt less holistic approaches of these concepts. Brittingham considers that the "province" concept is ambiguous and open to interpretation, as only a few States actually contribute to outer space practice, without sharing the benefits.[8] At best, "Province of mankind" should give access to space resources at a fair market price, "for

[6] A. Yokaris, International Air and Space Law (in Greek), Ant. N. Sakkoulas, Athens, 1996, p. 268–269.

[7] Idem.

[8] B.C. Brittingham, "Does the World Really Need New Space Law?", Oregon R.I.L., vol. 12, 2010, p. 37.

everyone",[9] whereas "Common heritage of mankind" should just impose a space exploration and exploitation for peaceful purposes.[10]

Tronchetti, for its part, draws a concrete distinction between the two concepts:

> Under the 'province of all mankind' concept, States are free to explore and use outer space as long as they do not harm other States. Although the exploration and use of outer space shall be carried out for the benefit and in the interests of all States, the 'province of all mankind' concept does not establish any obligation to share the benefits derived from outer space activities. By contrast, under the 'Common Heritage of Mankind' concept the exploration and exploitation of a certain 'area' and its resources shall be carried out in accordance with the rules established by an international regime or authority. Successful explorers, users and exploiters, would be obliged to conform to that international regime and share the benefits derived from their exploitative activities. In such sharing of the benefits, particular regard must be paid to the developing States, regardless of their level of participation in such activities.[11]

Finally, Gabrynowicz considers that a "strategic distinction" exists between the two concepts, as "the 'province of all mankind' provision contained in the Outer Space Treaty refers to 'activities (exploration and use)'", while "the 'common heritage' provision as contained in the Moon Treaty refers to 'material objects'".[12]

In order to reconcile opposing views on the matter, Goedhuis proposed, as a common denominator, four essential elements of the "Heritage" notion: that (1) the area under consideration cannot be subject to appropriation; (2) all countries must share in its management; (3) there must be an active sharing of benefits reaped from the exploitation of these resources; and (4) the area must be utilized exclusively for peaceful purposes.[13] However, one could observe that a reference to "sharing" could not be avoided.

***

In any case, the cornerstone for the "Province" and "Heritage" concepts is the "*common interest of mankind*" provision: According to the Preamble of the Outer Space Treaty, there is a "common interest of all mankind" in the progress of use of outer space, whereas such "use" "should be *carried on for the benefit of all peoples, irrespective of the degree of their economic or scientific development*". This "common interest of mankind" in outer space had already been recognized in the "historic" UNGA Resolutions 1348(XIII) of 13 December 1958, 1472(XIV) of 12 December

[9] Brittingham, *op. cit.*, p. 39.

[10] J.L. Zell, "Putting a Mine on the Moon: Creating an International Authority to Regulate Mining Rights in Outer Space", Minnesota J.I.L., vol. 15, 2006, p. 496.

[11] F. Tronchetti, The Exploitation of National Resources of the Moon and Other Celestial Bodies, Nijhoff, 2009, p. 44–45.

[12] J.I. Gabrynowicz, "The 'province' and 'heritage' of mankind reconsidered: A new beginning", NASA Johnson Space Center, 2nd Conference on Lunar Bases and Space Activities of the 21st century, vol. 2, 1992, p. 692.

[13] See D. Goedhuis, "Some Recent Trends in the Interpretation and Implementation of the Rules of International Space Law", Columbia J.T.L., vol. 19, 1981, p. 212.

1959, 1721(XVI) of 20 December 1961, 1962(XVIII) of 13 December 1963, and, relatively recently, 55/122 of 27 February 2001 where a desire for "the energetic promotion" of the "fullest… exploitation of outer space for the benefit of mankind" had also been expressed.[14] Moreover, the first of the famous "principles" contained in Resolution 1962/1963 provides that "…use of outer space shall be carried on for the benefit and in the interests of all mankind". The third principle also provides that "outer space and celestial bodies are not subject to national appropriation by claim of sovereignty, by means of use or occupation, or by any other means". In particular, Resolution 1962/1963, in reaffirming previous resolutions, provides, in its Preamble, that *«…Use of outer space should be carried on for the betterment of mankind and for the benefit of States irrespective of their degree of economic or scientific development*".

Reference must also be made to the Declaration on International Cooperation in the Exploration and Use of Outer Space for the Benefit and in the Interest of All States, Taking into Particular Account the Needs of Developing Countries (UNGA Resolution A/RES/51/122 of 13 December 1996). According to one of the included principles, "All States, particularly those with relevant space capabilities and with programmes for the exploration and use of outer space, should contribute to promoting and fostering international cooperation on an equitable and mutually acceptable basis. In this context, particular attention should be given to the benefit and the interests of developing countries and countries with incipient space programmes stemming from such international cooperation conducted with countries with more advanced space capabilities". Nevertheless, as Jakhu has pointed out, the "international cooperation" mentioned in this resolution is far from taking a mandatory form, given that, according to another of the included principles, "States are free to determine all aspects of their participation in international cooperation in the exploration and use of outer space on an equitable and mutually acceptable basis".[15] It is indeed remarkable that in the case of A/RES/51/122 the focus is on the "benefit and interest of all *States*" rather than on "mankind".

In respect of the "Common heritage of mankind" provision, it would be useful to go back to the origins of its conception. The relative term first appears in a proposal of Malta to the Secretary General of the United Nations, through its representative Arvid Pardo: By note verbale dated 18 August 1967, Malta proposed the inclusion in the agenda of the 22nd Session of the UN General Assembly the adoption of a Declaration concerning the international regulation of the seabed and the ocean floor. In a historic statement before the First Committee of the United Nations, on November 1st, 1967, Pardo, *inter alia*, stated:

> In the light of current technological developments, however, the compromise turns out to be no compromise at all; it is clear that the sea-bed beyond the 200-metre isobath will soon be subject to exploitation. The only question is, will it be exploited under national auspices for

[14] See also Resolution 1348(XIII) of 13 December 1958 (cited above).

[15] R. Jakhu, "United Nations Principles in Outer Space", in Proceedings of the United Nations/Nigeria Workshop on Space Law on "Meeting international responsibilities and addressing domestic needs" held on 21–24 November 2005, in Abuja, Nigeria, pp. 28–38.

> *national* purposes, *or will it be exploited under international auspices and for the benefit of mankind*?... The sea-bed and the ocean floor are a *common heritage of mankind* and should be used and exploited for peaceful purposes and for the exclusive benefit of mankind *as a whole*.[16]

It is certainly a fact that the industrialized countries have shown no intention to share the aspirations of Arvid Pardo, both with respect to the ocean depths as well as in relation to the Moon and other celestial bodies.[17] This reluctance, derived from their technological superiority, dictated to them an approach of the "common heritage of mankind" concept which essentially meant "all States shall have access to the outer space resources (although, for some of them, the Moon is too far)".[18] Thus, the dispute over the interpretation of the "Common Heritage" doctrine led to the formation of the following question: equitable sharing (of benefits) or equal access (to space resources)?[19]

***

Through this debate, the fundamental concept is finally put forward:

Whether outer space is considered 'the province of all mankind' or 'the common heritage of mankind', in any case, the dominant conceptual element remains the same: It is '(hu)mankind' that underpins both concepts.

Humankind, as a notion, is based on consistency, not on division: It is thus a profoundly different concept from "every nation". According to Baslar, "humankind" was generally associated with the notion of "all States", while, according to another view, the term refers to "all peoples".[20] Gorove also emphasized this sense of community, when he considered that humankind describes "a collective body of peoples wherever they may be found".[21]

It has already been proposed that humankind constitutes a distinct (from States) subject of international space law. Although States (at least some of them) have attempted to overlook the collective element inherent in the term '(hu)mankind'—especially since its reference in the Outer Space Treaty coincides with Pardo's monumental speech before the First Committee in 1967—its inclusion in existing international legal instruments has hindered the acceptance of private business

---

[16] Statement of Arvid Pardo, 1 November 1967, First Committee, UNGA, 22 UN GAOR, 1515th & 1516th Meeting, p. 8–9 (1515th) and 2 (1516th) (emphasis by the author).

[17] K.M. Zullo, "The Need to Clarify the Status of Property Rights in International Space Law", Georgetown L.J., vol. 90. 2001–2002, p. 2424.

[18] See R.J. Lee, Law and Regulation of Commercial Mining of Minerals in Outer Space, Springer, 2012, p. 15.

[19] D.A. Porras, "The 'Common Heritage' of Outer Space: Equal Benefits for most of Mankind", California West. I.L.J., vol. 37, no 1, 2006, p. 154.

[20] K. Baslar, The Concept of the Common Heritage of Mankind in International Law, Nijhoff, 1998, p. 73; L.B. Malagar & M.A. Magdosa-Malagar, "International Law of Outer Space and the Protection of Intellectual Property Rights", Boston U.I.L.J., vol. 17, 1999, p. 343.

[21] S. Gorove, "The Concept of the Common Heritage of Mankind: A Political, Moral, and Legal Innovation?", San Diego L.R., vol. 9, 1972, p. 393.

activities in space. It is evident that such views exacerbate the already unclear legal landscape.

2.1.2 The 1979 Moon Agreement

It is first important to stress that the 1979 Moon Agreement has a broader scope, which includes both orbits around the Moon and other celestial bodies within the Solar System. According to Article 1,

Article 1

1. The provisions of this Agreement relating to the Moon shall also apply to other celestial bodies within the solar system, other than the Earth, except insofar as specific legal norms enter into force with respect to any of these celestial bodies.
2. For the purposes of this Agreement reference to the Moon shall include orbits around or other trajectories to or around it.

…

Furthermore, Article 2 of the Moon Agreement reiterates the provision of Article III OST (application of international law to lunar activities), while Article 4 echoes Article I para. 1 OST (province of Mankind), however, making the interesting subordination of the exploration and use of the Moon "to the interests of present and future generations" (para. 1) and replicating the obligation of cooperation and mutual assistance of Article XI OST, giving it a special emphasis:

Article 2

All activities on the Moon, including its exploration and use, shall be carried out in accordance with international law, in particular the Charter of the United Nations, and taking into account the Declaration on Principles of International Law concerning Friendly Relations and Cooperation among States in accordance with the Charter of the United Nations,6 adopted by the General Assembly on 24 October 1970, in the interest of maintaining international peace and security and promoting international cooperation and mutual understanding, and with due regard to the corresponding interests of all other States Parties.

…

Article 4

1. The exploration and use of the Moon shall be the province of all mankind and shall be carried out for the benefit and in the interests of all countries, irrespective of their degree of economic or scientific development. Due regard shall be paid to the interests of present and future generations as well as to the need to promote higher standards of living and conditions of economic and social progress and development in accordance with the Charter of the United Nations.
2. States Parties shall be guided by the principle of cooperation and mutual assistance in all their activities concerning the exploration and use of the Moon. International cooperation in pursuance of this Agreement should be

as wide as possible and may take place on a multilateral basis, on a bilateral basis or through international intergovernmental organizations.

Further: Article 5 para. 1 of the Moon Agreement follows Article XI OST in requiring that the Secretary-General of the United Nations be informed of the activities of States relating to the exploration and use of the Moon:

Article 5

1. States Parties shall inform the Secretary-General of the United Nations as well as the public and the international scientific community, to the greatest extent feasible and practicable, of their activities concerned with the exploration and use of the Moon.

Article 7 (para. 1) introduces, as a limit to the use and exploration of the Moon by States, the protection of the lunar environment, thus echoing Article IX OST:

Article 7

1. In exploring and using the Moon, States Parties shall take measures to prevent the disruption of the existing balance of its environment, whether by introducing adverse changes in that environment, by its harmful contamination through the introduction of extra-environmental matter or otherwise. States Parties shall also take measures to avoid harmfully affecting the environment of the Earth through the introduction of extraterrestrial matter or otherwise.

***

As mentioned, there is no specific, *ad hoc* legal regime governing the administration of the exploration, exploitation, and use of space resources. However, the 1979 Moon Agreement, despite its weak ratification, establishes such a regime, as provided for in Article 11:

Article 11

1. The Moon and its natural resources are the common heritage of mankind, which finds its expression in the provisions of this Agreement, in particular in paragraph 5 of this article.
2. The Moon is not subject to national appropriation by any claim of sovereignty, by means of use or occupation, or by any other means.
3. Neither the surface nor the subsurface of the Moon, nor any part thereof or natural resources in place, shall become property of any State, international intergovernmental or non-governmental organization, national organization or non-governmental entity or of any natural person. The placement of personnel, space vehicles, equipment, facilities, stations and installations on or below the surface of the Moon, including structures connected with its surface or sub- surface, shall not create a right of ownership over the surface or the subsurface of the Moon or any areas thereof. The foregoing provisions are without prejudice to the international regime referred to in paragraph 5 of this article.

4. States Parties have the right to exploration and use of the Moon without discrimination of any kind, on the basis of equality and in accordance with international law and the terms of this Agreement.
5. States Parties to this Agreement hereby undertake to establish an international regime, including appropriate procedures, to govern the exploitation of the natural resources of the Moon as such exploitation is about to become feasible. This provision shall be implemented in accordance with article 18 of this Agreement.
6. In order to facilitate the establishment of the international regime referred to in paragraph 5 of this article, States Parties shall inform the Secretary-General of the United Nations as well as the public and the international scientific community, to the greatest extent feasible and practicable, of any natural resources they may discover on the Moon.
7. The main purposes of the international regime to be established shall include:

    (a) The orderly and safe development of the natural resources of the Moon;
    (b) The rational management of those resources;
    (c) The expansion of opportunities in the use of those resources;
    (d) An equitable sharing by all States Parties in the benefits derived from those resources, whereby the interests and needs of the developing countries, as well as the efforts of those countries which have contributed either directly or indirectly to the exploration of the Moon, shall be given special consideration.

8. All the activities with respect to the natural resources of the Moon shall be carried out in a manner compatible with the purposes specified in paragraph 7 of this article and the provisions of article 6, paragraph 2, of this Agreement.

Article 8 is also of relevance:

Article 8

1. States Parties may pursue their activities in the exploration and use of the Moon anywhere on or below its surface, subject to the provisions of this Agreement.
2. For these purposes States Parties may, in particular:

    (a) Land their space objects on the Moon and launch them from the Moon;
    (b) Place their personnel, space vehicles, equipment, facilities, stations and installations anywhere on or below the surface of the Moon.

Personnel, space vehicles, equipment, facilities, stations and installations may move or be moved freely over or below the surface of the Moon.

3. Activities of States Parties in accordance with paragraphs 1 and 2 of this article shall not interfere with the activities of other States Parties on the Moon. Where such interference may occur, the States Parties concerned shall undertake consultations in accordance with article 15, paragraphs 2 and 3, of this Agreement.

However, it is obvious that, for the majority of States that have not yet ratified this Agreement, the applicable international legal rules, in relation to the exploration and exploitation of outer space, are the aforementioned provisions of the OST.

In any case, it remains that the Moon Agreement advocates for a collective regime for the exploitation of space resources of the Moon (as well as of other celestial bodies within the Solar System), at least at the level of principle.

2.2 Inadequacies of National Laws

2.2.1 In General

International agreements represent by large one of the main sources of international law according to Article 38 para. 2 literal (b) of the ICJ Statute.[22] In this sense, the customary rule *pacta sunt servanda*, which is also a fundamental principle of every national legal system, is one of major importance for international law.[23] In fact the performance of a treaty may reveal the existence of legal gaps or ambiguous terms that allow States to adopt a different—even contradictory—approach as to the exact content of their contractual obligations. In order to resolve such disputes, States parties to a treaty may have recourse to the customary rules of treaty interpretation[24] codified in the 1969 Vienna Convention on the Law of Treaties.[25]

Particularly, based on the general rules of treaty interpretation set forth in Articles 31 and 32 VCLT, State practice, which is subsequent to the conclusion of an international treaty, may constitute a useful supplementary means for the interpretation of treaty obligations. In more detail, according to Article 31 para. 3 literal (b) VCLT "*There shall be taken into account, together with the context: [...] any subsequent practice in the application of the treaty which establishes the agreement of the parties regarding its interpretation;*". Said provision requires active practice of some parties to the treaty, which should further be consistent rather than haphazard and should have occurred with a certain frequency.[26] However, the subsequent practice must establish "the agreement of the parties regarding its interpretation", in the sense that it will have been acquiesced in by the other parties and no other party will

[22] Statute of the International Court of Justice, article 38, para. 1.

[23] M. Lachs, "Pacta Sunt Servanda", in: 7 ENCYCLOPEDIA OF PUBLIC INTERNATIONAL LAW 368; S. Reinhold, Good Faith in International Law, 2 UCLJLJ, 61 (2013).

[24] Maritime Dispute (Peru v. Chile) (Judgment) 2014 I.C.J. 28 (Jan. 27); Sovereignty over Pulau Ligitan and Pulau Sipadan (Indonesia v. Malaysia) (Judgment) 2002 I.C.J. 645 (Dec. 17); Kasikili/ Sedudu Island (Botswana v. Namibia) (Judgement) 1999 I.C.J. 1059 (Dec. 13).

[25] Vienna Convention on the Law of Treaties, done at Vienna on 23 May 1969, entered into force on 27 January 1980, 1155 U.N.T.S. 331 [hereinafter VCLT].

[26] M.E. Villiger, Commentary on the 1969 Vienna Convention on the Law of Treaties, 2009, p. 432–433.

have raised an objection.[27] Nonetheless, if these conditions are not met, such practice may still serve as a supplementary means of interpretation according to Article 32 VCLT,[28] In this light, the term "State practice" may encompass *inter alia* the national legislations unilaterally adopted by the States parties to an international treaty.

Furthermore, State practice along with *opinio juris* constitutes one of the fundamental elements for the establishment of customary rules, which also constitute a source of international law under Article 38 para. 1 literal (b) of the ICJ Statute, as evidence of general practice of States accepted as law. Besides, international custom and international conventions are two different sources of international law that may be formed and applied equally and in parallel, as acknowledged by the International Court of Justice in the *Nicaragua* case.[29]

***

In terms of State practice, up to date only four States parties to the OST have enacted national legislations that explicitly provide their nationals with the right to conduct space mining activities for purposes of commercial exploitation, namely the United States in 2015, Luxembourg in 2017, the United Arab Emirates in 2019 and Japan in 2022. These national laws are set out below.

2.2.2 *Existing National Legislations*

2.2.2.1 United States: The 2015 U.S. Commercial Space Launch Competitiveness Act

In the U.S., commercialization of outer space was already encouraged by the *U.S. Commercial Space Launch Act* of 1984, which states that "the general welfare of the United States requires that the Administration seek and encourage, to the maximum extent possible, the fullest commercial use of space". At that time, the Act was essentially concerned with unmanned activities, as the issue of human spaceflight was not yet raised. In addition, the Act authorised the Federal Aviation Administration (FAA) (through a grant of authority) to encourage, facilitate and promote commercial space launches and re-entries by the private sector (§ 50,903).

The United States, which is a major space faring State and one of the most important pioneers in the space domain, was the first State to adopt a national legislation that explicitly provided for the right to undertake space mining activities for

[27] *Idem.*

[28] S.T. BERNÁRDEZ, "(Chapter) 37 - Interpretation of Treaties by the International Court Of Justice Following the Adoption of the 1969 Vienna Convention on the Law of Treaties", Liber Amicorum Professor Ignaz Seidl-Hohenveldern, Brill, 1998, p. 726.

[29] ICJ, Military and Paramilitary Activities in and against Nicaragua (Nicaragua v. United States of America), Merits, Judgment. I.C.J. Reports 1986, p. 14, at p. 93.

commercial purposes. On November 25, 2015, the President of the United States signed the *U.S. Commercial Space Launch Competitiveness Act (H.R. 2262)*. The purpose of said Act is to "To facilitate a pro-growth environment for the developing commercial space industry by encouraging private sector investment and creating more stable and predictable regulatory conditions, and for other purposes". With respect to space mining, of particular interest is Title IV—Space Resource Exploration and Utilization. This chapter emphasizes the promotion of the right of American citizens "to engage in commercial exploration for and commercial recovery of space resources free from harmful interference, in accordance with such obligations and subject to authorization and continuing supervision by the federal government". In addition, "A U.S. citizen engaged in commercial recovery of an asteroid resource or a space resource shall be entitled to any asteroid resource or space resource obtained, including to possess, own, transport, use, and sell it according to applicable law, including U.S. international obligations".

In particular:

According to Chapter 513 para. 51302 literals (a) of the 2015 U.S. Commercial Space Launch Competitiveness Act,

> IN GENERAL.—The President, acting through appropriate Federal agencies, shall— "(1) facilitate commercial exploration for and commercial recovery[30] of space resources by United States citizens; (2) discourage government barriers to the development in the United States of economically viable, safe, and stable industries for commercial exploration for and commercial recovery of space resources in manners consistent with the international obligations of the United States; and (3) promote the right of United States citizens to engage in commercial exploration for and commercial recovery of space resources free from harmful interference, in accordance with the international obligations of the United States and subject to authorization and continuing supervision by the Federal Government".

What is more, para. 51303 of the aforementioned Chapter entails an additional provision that deals separately with the mining of asteroids, pursuant to which:

> A United States citizen engaged in commercial recovery of an asteroid resource or a space resource under this chapter shall be entitled to any asteroid resource or space resource obtained, including to possess, own, transport, use, and sell the asteroid resource or space resource obtained in accordance with applicable law, including the international obligations of the United States.

Finally, Section 403 of the Act assures that the United States does not assert sovereignty or sovereign or exclusive rights or jurisdiction over, or the ownership of, any celestial body.

Consistent with the aforementioned Act, U.S. space policy on space mining has gradually shifted from a science-driven approach to one increasingly focused on commercial objectives. Indicatively, in the 2010 space policy directive titled

[30] See L.I. Tennen's reflection on the concept of commercial recovery, *in:* "Enterprise Rights in Extraterrestrial Resources: Commentary on Outer Space and International Geography: Article II and the Shape of Global Order", New England Law Review, Vol. 52, No. 2, Winter 2018, p. 144 *et seq.*

"National and Commercial Space Programs",[31] the mineral resources of celestial bodies are referred to as "scientific data", which under this qualification can be the subject of lawful acquisition.

> § 50113. Acquisition of space science data (a) DEFINITION OF SPACE SCIENCE DATA.—In this section, the term "space science data" includes scientific data concerning—(1) the elemental and mineralogical resources of the moon, asteroids, planets and their moons, and comets;.

Moreover, under the title "Civil Space Guidelines Space Science, Exploration, and Discovery" of the same directive it is provided that:

> The Administrator of NASA shall: [...] Pursue capabilities, in cooperation with other departments, agencies, and commercial partners, to detect, track, catalog, and characterize near-Earth objects to reduce the risk of harm to humans from an unexpected impact on our planet and to identify potentially resource-rich planetary objects.

Hence, in the past the U.S. promoted space minerals through their policy and legislation as a scientific data subject to research, exploration and use.

At the other end of this transition lies the 2015 Commercial Space Launch Competitiveness Act (in other words, through a unilateral act of State), under which U.S. citizens are expressly entitled, *inter alia*, to "possess", "own", and "sell" mineral resources from asteroids and other celestial bodies. Thus, the Act expressly grants U.S. citizens property rights over the mineral resources of outer space, a concession that seems to be in contradiction to the principle of non-appropriation as enshrined in Article II OST.

In relation to this Act, arguments have been developed, in theory, in favor of the permissibility of the exploitation/appropriation of the mineral resources of outer space. However, the prevailing view on the exact interpretation of Article II OST is that the principle of non-appropriation prohibits both the exercise of sovereign rights over celestial bodies (by States) and the right of ownership by non-State actors.[32] This conclusion is further reinforced by the language of Article 11 para. 3 of the Moon Agreement, according to which "Neither the surface nor the subsurface of the Moon, nor any part thereof or natural resources in place, shall become property of any State, international intergovernmental or non-governmental organization, national organization or non-governmental entity or of any natural person".

In this regard, the International Institute of Space Law (IISL) adopted an intermediate position. The Institute, in its *Position Paper on Space Resource Mining,*[33] after having recalled that Article 11 paragraph 3 of the Moon Agreement of 1979

---

[31] Government of the United States of America, National Space Policy of the United States of America, § 50,113 (June 28th 2010), available at https://history.nasa.gov/national_space_policy_6-28-10.pdf (last accessed: 31.12.2024).

[32] S. Freeland & R. Jakhu, "Article II", *in:* S. Hobe, B. Schmidt-Tedd, K.-U. Schrogl & G. Meishan Goh (eds.), Cologne Commentary on Space Law, Vol. 1 Outer Space Treaty, Carl Heymanns Verlag, 2009, p. 50.

[33] IISL, Position Paper on Space Resource Mining, adopted by consensus by the Board of Directors on 20 December 2015, online: http://www.iislweb.org/html/20151220_news.html (Last accessed: 20.05.2016).

was not binding upon the United States and "has not gained the status of a rule of customary international law", further mentioned that "in view of the absence of a clear prohibition of the taking of resources in the Outer Space Treaty one can conclude that the use of space resources is permitted. Viewed from this perspective, the new United States Act is a possible interpretation of the Outer Space Treaty. Whether and to what extent this interpretation is shared by other States remains to be seen". Thus, the IISL had recourse, in a way, to the so-called "*Lotus* principle", which can be summarized in the *dictum* "Everything which is not forbidden is allowed".[34] According to other views, the U.S. Space Act directly violates the Outer Space Treaty, as it allows States, private firms or international organizations to appropriate natural space resources.[35]

It is worth mentioning that the issue of claiming property rights in outer space by individuals was raised before the US courts in the past, in the context of the *Nemitz* case.[36] Particularly, on February 13th, 2003, Gregory Nemitz, a US citizen (CEO of Orbital Development of San Diego, a company based in the United States) submitted a request to NASA to pay him a $20 fine for parking Shoemaker's ship on the asteroid 433 Eros. Nemitz claimed that he owned the aforementioned asteroid (!) following a filing in 2000 as part of a claim filed in 2000 by an organization called Archimedes Institute. However, NASA deemed Nemitz's request as legally invalid, since, as it reasoned, Article II OST prohibited ownership of asteroids by any private party. In its letter to Nemitz dated 9 April 2001, NASA clearly mentioned, *inter alia*, that "Your individual claim of appropriation of a celestial body (the asteroid 433 Eros) appears to have no foundation in law".[37] The US Department of State also rejected Nemitz's claims, pointing out that "private ownership of an asteroid is precluded by Article II of the [Outer Space Treaty]" (reply to Nemitz dated 15 August 2003).[38] By the same reasoning, the Federal Court of the State of Nevada and, later,

[34] PCIJ, The Case of the SS 'Lotus', Series A No 10, 7 September 1927, p. 4. For the Lotus case, see Ph. Pazartzis, "Judicial Activism and Judicial Self-Restraint: The PCIJ's Lotus Case" *in:* Ch.J. Tams, M. Fitzmaurice & P. Merkouris (eds.), Legacies of the Permanent Court of International Justice (Nijhoff, 2013), p. 319–335, A. Pellet, "Lotus que des sottises on profère en ton nom! Remarques sue le concept de souveraineté dans la jurisprudence de la Cour mondiale", *in:* Mélanges en l'honneur de Jean-Pierre Puissochet, L'Etat souverain dans le monde d'aujourd'hui, Pedone, 2008, at p. 215. See also G.D. Kyriakopoulos, «Legal Challenges Posed by the Action of Non-State Actors in Outer Space», *in:* McGill Centre for Research in Air and Space Law, M. Manoli & S.B. Habchi (eds.), Monograph Series V – Conflicts in Space and the Rule of Law, June 2017, p. 273–290, at p. 285.

[35] Katrina Pascual, "U.S. Space Mining Law Is Potentially Dangerous And Illegal: How Asteroid Mining Act May Violate International Treaty" (28 November 2015) online:
<http://www.techtimes.com/articles/111534/20151128/u-s-space-mining-law-is-potentially-dangerous-and-illegal-how-asteroid-mining-act-may-violate-international-treaty.htm> (Last accessed: 20.05.2016).

[36] For the Nemitz case, see L.I. Tennen, "Towards a New Regime for Exploitation of Outer Space Mineral Resources", Nebraska Law Review, Vol. 88, No. 4, 2010, p. 809 *et seq.*

[37] See https://www.orbdev.com/010409.html (last accessed: 01.03.2025).

[38] https://www.erosproject.com/exhibit01.html (last accessed: 01.03.2025).

the Court of Appeals, dismissed Nemitz's lawsuit in this regard as legally invalid.[39] It can be said that this reaction to Nemitz's claims, both by public authorities and U.S. courts, which obviously constitute an official practice of the United States, clearly indicate the adoption of a broad interpretation of Article II, which was made in due course.

On 6.4.2020, the *Executive Order on Encouraging International Support for the Recovery and Use of Space Resources* was issued by U.S. President Donald Trump, as part of the aforementioned Commercial Space Launch Competitiveness Act.[40] In this order, the Moon Agreement of 1979 is completely renounced:

> ...the United States does not consider the Moon Agreement to be an effective or necessary instrument to guide nation states regarding the promotion of commercial participation in the long-term exploration, scientific discovery, and use of the Moon, Mars, or other celestial bodies. Accordingly, the Secretary of State shall object to any attempt by any other state or international organization to treat the Moon Agreement as reflecting or otherwise expressing customary international law (Section 2).

…while for the first time an American legislative instrument contains an explicit denial of the global commons character of outer space: "Outer space is a legally and physically unique domain of human activity, and the United States does not view it as a global commons".

Section 3 of the Order instructs the US Secretary of State, in cooperation with appropriate Secretaries and NASA (Secretary of Commerce, Secretary of Transportation, Administrator of the National Aeronautics and Space Administration) to act "to encourage international support for the public and private recovery and use of resources in outer space... In carrying out this section, the Secretary of State shall seek to negotiate joint statements and bilateral and multilateral arrangements with foreign states regarding safe and sustainable operations for the public and private recovery and use of space resources".

The ambiguous language of Section 3 could be seen as revealing a regulatory inconsistency—an attempt to reconcile current U.S. ambitions regarding the exploitation of space resources with both the spirit and letter of the existing international legal framework.

Last but not least: In contrast with the space policy of the last decade, the U.S. space policy directive of 2020—[41]in consistence with the 2015 Space Launch Competitiveness Act—deals with mineral resources of outer space not only as an object of scientific research, but mainly as a source of commercial profit. For instance, among the Principles provided under the 2020 Policy Directive it is expressly declared that "*The United States will pursue the extraction and utilization*

[39] US DISTRICT COURT, DISTRICT OF NEVADA, Nemitz v. United States of America et al., Dismissal of Plaintiff's claim, with prejudice, of 26 April 2004.

[40] Text in: https://trumpwhitehouse.archives.gov/presidential-actions/executive-order-encouraging-international-support-recovery-use-space-resources/ (last accessed: 04.04.2023).

[41] Government of the United States of America, National Space Policy of the United States of America, p. 3 (December 9th 2020), available at https://trumpwhitehouse.archives.gov/wp-content/uploads/2020/12/National-Space-Policy.pdf (last accessed: 29.03.2023).

of space resources in compliance with applicable law, recognizing those resources as critical for sustainable exploration, scientific discovery, and commercial operations".

2.2.2.2 Luxembourg: Loi du 20 juillet 2017 sur l'exploration et l'utilisation des ressources de l'espace

Luxembourg adopted a law on space exploration and the use of space resources in 2017 (Loi du 20 juillet 2017 sur l'exploration et l'utilisation des ressources de l'espace). Article 1 of this law states that "space resources are capable of being appropriated", which means that future extraction companies will acquire ownership rights over the space resources they extract. In addition, according to Articles 2–4 of the Law, Luxembourg companies or European companies established in Luxembourg may extract space mineral resources for commercial use after obtaining approval from the Luxembourg Government. Again, the compatibility of these provisions with the principle of non-appropriation is highly questionable.

2.2.2.3 United Arab Emirates: The U.A.E. Federal Law on the Regulation of the Space Sector (2019)

The U.A.E. have entered the space arena with an ambitious space program that includes space mining for commercial purposes. The U.A.E. which is currently a major supplier of fossil fuel on Earth, appears to be keen in preserving its advantage at the energy market by extending its interest to the minerals of the moon and other celestial bodies. In this light the U.A.E. adopted its Federal Law No. (12) of 2019 "On the Regulation of the Space Sector".[42] According to Article 4(i) and (ii) of the U.A.E. Federal Law, the latter regulated *inter alia* space resources exploration or extraction activities, as well as activities for the exploitation and use of Space Resources for scientific, commercial, or other purposes. Furthermore, Article 18 of the U.A.E. Federal Law titled "*Exploration, Exploitation and Use of Space Resources*" provides that:

> Subject to the provisions of Article (14) of this Law, the conditions and controls relating to Permits for the exploration, exploitation and use of Space Resources, including their acquisition, purchase, sale, trade, transportation, storage and any Space Activities aimed at providing logistical services in this regard shall be determined by a decision issued by the Council of Ministers or whomever it delegates. 2- The Permits referred to in clause (1) of this Article shall be granted by a decision of the Board of Directors upon the proposal of the Director General.

[42] Government of the United Arab Emirates, Federal Law No. (12) of 2019 Issued on 19/12/2019 Corresponding to 22 Rabi' Al-Akhar 1441H. ON THE REGULATION OF THE SPACE SECTOR, art. 4 (December 19th 2019).

In the same context, the U.A.E. have also issued a national space policy, stating that "*5.1.6. Support to Natural Resources Management Natural resources exploration, utilization, investment, and preservation are critical to the economy of the UAE*".[43] Accordingly, the space agenda of the UAE includes the exploration of new natural resources, including new gas and petroleum reserves, and support to their extraction, as well as exploration, mining, extraction, and utilization of resources in space.

2.2.2.4 Japan: The Japanese Act on the Promotion of Business Activities Related to the Exploration and Development of Space Resources (2021)

Finally, Japan was the latest space-faring nation to adopt a national legislation on space mining. Specifically, on 15 June 2021, the Japanese Diet enacted the "Act on the Promotion of Business Activities Related to the Exploration and Development of Space Resources" (the Space Resources Act),[44] which permits persons to conduct business activities for the exploration and development of space resources and to acquire ownership of space resources.[45] In this connection, it is interesting to note that the said act provides among others for a detailed regime of registration of activities concerning the extraction and exploitation of space resources.

2.2.2.5 An Appraisal of the Impact of National Legislation on the Regulation of Space Resources

- **The relevant legislative practice is far from being uniform and general.** What follows from the above is that over the past few years certain space-faring States have commenced an effort to overcome legal issues pertaining to the interpretation of obligations enshrined in Article II OST through legislative initiatives at national level. On this note, although, in international law, State practice, including national legislation as well, may serve a means of treaty interpretation as per Articles 31 and 32 VCLT, such subsequent State practice must be uniform and not objected by other States Parties.[46] Similarly, for the establishment of

[43] Government of the United Arab Emirates, National Space policy of the United Arab Emirates, §§ 5.1.6, 2.2., 5.1.6., 5.2.4., 5.2.5., 5.3.5 (2019).

[44] Government of Japan, Act on the Promotion of Business Activities Related to the Exploration and Development of Space Resources, The Space Resources Act, Law No. 83 of 2021, art. 2, 3 (June 15th 2021).

[45] Hiroko Yotsumoto et al., Introduction to the national legal, regulatory and policy framework, *in:* The Space Law Review: Japan, available at https://thelawreviews.co.uk/title/the-space-law-review/japan (last accessed: 23.03.2023).

[46] In this respect, see M. Manoli, The Immaterial Territoriality of Outer Space: Reconstructing Territorylessness in Modern Space Governance (Montreal: McGill University, 2021), online: EScholarship https://escholarship.mcgill.ca/concern/theses/hd76s530m, p. 191, as well as the relevant developments under Part V.

international custom as well, the criterion of "general State practice", as per Article 38 para. 1 literal (b) of the ICJ Statute, is interpreted as practice that has occurred in such a way as to show a general recognition that a rule of law or legal obligation is involved.[47] Nonetheless, with regards to national space legislation concerning the extraction and exploitation of space resources, the elements of "general" and "uniform" State practice do not—at least yet—appear to be fulfilled. On the contrary, even the national jurisprudence of major space-faring States that have enacted such national legislations seems to contravene such regulations. For instance, in the context of the *Nemitz* case, mentioned above, NASA, the Department of State and the competent Courts of first and second instance dismissed the claims of an individual claiming property over the asteroid 433 Eros by reference to Article II OST, which precludes property rights of private persons in outer space.

Apart from the U.S. jurisprudence, a similar case has also been brought before the Chinese courts. Specifically, since 1980, a company called "Lunar Embassy" with the nationality of the People's Republic of China has been selling plots of land on the Moon for 298 yuan (*i.e.* $37 U.S.) per acre. The case was brought before the Beijing People's Court, which decided the case in 2007. The owner of the company Li Jie applied for an extension of his company's operating license, arguing that national legislation did not contain a relevant regulation prohibiting private ownership of land on the Moon. He described the operation of his company as a 'beautiful dream'. However, the Court, referring to Article II OST, ruled that the action of the company in question was unlawful as it contravened the international obligations of the State under the OST and gave rise to the international responsibility of the People's Republic of China.[48]

Similarly, in 2012, referring again to Canada's obligations under Article II OST, Canadian Judge Alain Michaud declared Quebec resident Sylvio Langevin a "contentious litigant" and barred him from filing a claim asserting ownership over nine planets, four moons of Jupiter and the empty space between celestial bodies.[49]

Accordingly, it may be inferred from the above references that State practice in the area of space resource extraction and exploitation remains largely undefined and internally inconsistent. This conclusion is reinforced by the efforts of States at the international level to agree on a detailed legal regime for the regulation of lunar mining, which indicates that the majority of States remain hesitant to accept as lawful mining activities on the moon and other celestial bodies which are not conducted under an internationally accepted legal regime. In conclusion, unilateral State initiatives at the national level cannot prevail over internationally accepted rules of law

[47] ICJ, North Sea Continental Shelf, Judgment, I.C.J. Reports 1969, p. 3.

[48] Beijing authorities suspend license of "Lunar Embassy", available at http://www.chinadaily.com.cn/english/doc/2005-11/07/content_492152.htm (last accessed: 01.03.2025).

[49] Drew Hasselback, "The most-read Canadian court decision of 2012 is out of this world", 14.12.2012, available at https://financialpost.com/legal-post/the-most-read-canadian-court--decision-of-2012-is-out-of-this-world (last accessed: 01.03.2025).

contained in the OST, such as Article II, which have been further agreed by consensus. The former legislations however have the potential to create obligations of the legislative States towards their nationals and private entities of their nationality.

- **The applicability of international law with respect to activities in outer space.** Outer space belongs to the so-called *global commons*, which include spaces beyond national jurisdictions[50] (high seas, deep seabed, international airspace, Antarctica). This is clear, *inter alia*, in the combined reading of articles I paras. 1 and 2 OST: "The exploration and use of outer space... shall be carried out for the benefit and in the interests of all countries... and shall be the province of all mankind. What is more, "outer space... shall be free for exploration and use by all States without discrimination of any kind, on a basis of equality and in accordance with international law, and there shall be free access to all areas of celestial bodies". The absence of any national jurisdiction is also patent in the principle of non-appropriation, enshrined in article 2 OST, pursuant which "outer space, including the Moon and other celestial bodies, is not subject to national appropriation by claim of sovereignty, by means of use or occupation, or by" any other means".

Any activity in outer space is regulated by international law. This is apparent in the combined reading of articles I, III and VI of the OST. Although Article VI leaves room for "national" activities in outer space, Article III provides that "States Parties to the Treaty shall carry on activities in the exploration and use of outer space, including the Moon and other celestial bodies, *in accordance with international law, including the Charter of the United Nations,* in the interest of maintaining international peace and security and promoting international cooperation and understanding". The applicability of international law is further reinforced by Article I, where it is stipulated that "Outer space, including the Moon and other celestial bodies, shall be free for exploration and use by all States… *and in accordance with international law*". Further it must be mentioned that Article III OST sets a binding obligation,[51] which was universally accepted by the main space Powers of the 60s[52] during the drafting of the Outer Space Treaty.

With respect to the exploration and exploitation of outer space, international space law in force mainly comprises the Outer Space Treaty and the Moon Agreement. At present, 17 States have ratified the MOON which provides, in its Article 11, for an international regime of exploitation of the natural resources of the Moon, "as such exploitation is about to become feasible". For the majority of States that have not ratified said instrument, the international norms applicable to the exploration and exploitation of outer space are the relevant provisions of the Outer

[50] S. Ranganathan, "Global Commons", European Journal of International Law, Volume 27, Issue 3, 1 August 2016, Pages 693–717.

[51] See O. Ribbelink, "Article III", *in:* S. Hobe, B. Schmidt-Tedd & K-U. Schrogl, Cologne Commentary on Space Law, Vol. 1, Outer Space Treaty, Carl Heymanns Verlag, 2009, p. 66–67.

[52] USSR, UK and USA.

Space Treaty of 1967. In this context, particular attention must be given to the (aforementioned) articles I and II OST.

The applicability of international law with respect to activities in outer space is, at least implicitly, recognized by most national space legislations, which, in general, delimit their scope of application to matters that, although related to outer space, are closely linked with territorial sovereignty and state jurisdiction on Earth: For instance, the Belgian Space Law of 17 September 2005[53] and the Greek Law 4508/2017 mainly regulate topics such as the licensing of space activities, the transfer of activities, national registration of space objects as well as liabilities, counterclaims and measures in the event of falling space objects.[54] What is more, the primacy of international law is almost explicitly recognized, as articles 2 and 4 of the Belgian Law provide that its applicability is limited, *rationae loci*, to places under the jurisdiction and control of Belgium, whereas the applicability of international norms with respect to outer space activities, enshrined in Article III OST, is solemnly repeated:[55]

> Art. 2 §1. This law covers the activities of launching, flight operations and guidance of space objects carried out by natural or legal persons in the zones placed under the jurisdiction or control of the Belgian State or using installations, personal or real property, owned by the Belgian State or which are under its jurisdiction or its control.
>
> §2. When provided for under an international agreement, this law may apply to the activities referred to under indent 1 and carried out by natural or legal persons of Belgian nationality, irrespective of the location where such activities are carried out.
>
> …
>
> Art. 4 §3: The activities must be carried out in accordance with international law and, in particular, with the principles laid down in the Outer Space Treaty and the other treaties and agreements to which Belgium is a party.[56]

Consequently, activities in outer space, either public or private in nature, are governed by international law. In general, The governance of celestial bodies, as well as of interplanetary space, is of an international character.[57] **This international gover-**

[53] As revised by the Law of 1 December 2013 – see http://www.belspo.be/belspo/space/belaw_en.stm (last accessed: 31.07.2018).

[54] Belgian Law of 1 December 2013, articles 13–15 (idem). See also the articles 3, 10, 11, 13 and 17 of the Greek Law 4508/2017 which contains similar arrangements.

[55] But as we have already seen, the U.S. Commercial Space Launch Competitiveness Act (H.R. 2262) seems to go to a different direction, as "commercial" exploration and recovery of space resources is encouraged and promoted, whereas U.S. citizens are explicitly entitled, by Act of State, to "possess", "own" and "sell" asteroid resources; thus the U.S. Act explicitly confers to U.S. citizens property rights to resources in outer space—Commercial Space Launch Competitiveness Act, H.R.2262, 114th Congress (2015–2016), https://www.congress.gov/bill/114th-congress/house-bill/2262/text (Last accessed: 20 May 2016). On 25 November 2015, the Act was signed by the President and became Public Law No: 114-90 – see information online: https://www.congress.gov/bill/114th-congress/house-bill/2262/all-info (Last accessed: 20 May 2016) – see G.D. Kyriakopoulos, "Legal Challenges Posed by the Action of Non-State Actors in Outer Space", *op.cit.*, p. 283 *et seq.*

[56] See also Articles 1(2) and 3(3) of the Greek Law 4508/2017 (*idem*).

[57] The term "governance" is hereby used with the meaning of "the way that [outer space] is managed at the highest level, and the systems for doing this".

https://dictionary.cambridge.org/dictionary/english/governance (last accessed: 01.04.25).

nance of space activities naturally encompasses the exploration and exploitation of space resources. It is therefore necessary to establish an *international* legal framework in this area. This does not exclude the possibility of granting rights to States or private actors, but these will also be based on international regulation. After all, it will not be the first time that areas of international interest (and beyond national jurisdiction) have been regulated by international law. The regulation of international airspace on the basis of international standards adopted by the International Civil Aviation Organization (ICAO),[58] or the regulation of the international seabed through the combined application of Part XI of the United Nations Convention on the Law of the Sea and the relevant 1994 Agreement,[59] are two examples that immediately come to mind.

***

Although the necessary regulations for the exploration and use of space resources must derive from international law, the applicable international space law, in its current form and under the constraints of specific fundamental principles, fails to provide the regulatory adequacy required by the endeavor of space resource exploitation. As indicated above, this insufficiency is twofold: on the one hand, it arises from the principles themselves (the freedom of use and exploration, in combination with the principle of non-appropriation, framed within the broader context of international cooperation and multilateralism as required by Article IX of the Outer Space Treaty); on the other hand, it stems from the absence of a well-structured, detailed, and multifaceted legal framework (similar to that governing the international seabed) that would address the specific challenges and circumstances involved in the exploitation of mineral resources located far from Earth.

Given the structural weaknesses in the current legal framework—and the inadequacy of unilateral national regulations, for reasons already outlined—the question arises whether legal scholars can contribute to overcoming these deadlocks through interpretative insights, or whether this can only be achieved through coordinated action and cooperation by States on the international stage, allowing the current suspended step of international law to gradually evolve into a more secure and consistent one.

[58] On the basis of Article 12 of the 1944 Chicago Convention on International Civil Aviation: "... Over the high seas, the rules in force shall be those established under this Convention".

[59] For the management of international seabed resources, see *infra*, Part V.

Chapter 3
Uncertainty of the Next Step: The Limitations of Academic Approaches in Offering a Satisfactory Response

Previous monographs on the issue of the exploitation of space resources in general have dealt extensively with the admissibility of such activities under international law in force. Their analyses focus in particular on the relationship between Articles I (freedom of use and exploration, freedom of access, freedom of scientific research), II (principle of non-appropriation) and IX of the Outer Space Treaty (due regard, forward and backward contamination, consultations), in the light, of course, of Article VI of the Outer Space Treaty, which establishes the international responsibility of States also for "national" activities of non-governmental entities, for which States are required to grant "authorization" and exercise "continuing supervision". These provisions are indeed central to the overall debate on the legal dimension of the utilization and exploitation of space resources and are therefore set out here:

> Article I
>
> The exploration and use of outer space, including the Moon and other celestial bodies, shall be carried out for the benefit and in the interests of all countries, irrespective of their degree of economic or scientific development, and shall be the province of all mankind.
>
> Outer space, including the Moon and other celestial bodies, shall be free for exploration and use by all States without discrimination of any kind, on a basis of equality and in accordance with international law, and there shall be free access to all areas of celestial bodies.
>
> There shall be freedom of scientific investigation in outer space, including the Moon and other celestial bodies, and States shall facilitate and encourage international cooperation in such investigation.
>
> Article II
>
> Outer space, including the Moon and other celestial bodies, is not subject to national appropriation by claim of sovereignty, by means of use or occupation, or by any other means.
>
> Article VI
>
> States Parties to the Treaty shall bear international responsibility for national activities in outer space, including the Moon and other celestial bodies, whether such activi-

G. D. Kyriakopoulos, *Space Resource Utilization and Exploitation: The Suspended Step of International Law*, Space and Society,
https://doi.org/10.1007/978-3-031-96336-0_3

> ties are carried on by governmental agencies or by non-governmental entities, and for assuring that national activities are carried out in conformity with the provisions set forth in the present Treaty. The activities of non-governmental entities in outer space, including the Moon and other celestial bodies, shall require authorization and continuing supervision by the appropriate State Party to the Treaty. When activities are carried on in outer space, including the Moon and other celestial bodies, by an international organization, responsibility for compliance with this Treaty shall be borne both by the international organization and by the States Parties to the Treaty participating in such organization.
>
> Article IX
>
> In the exploration and use of outer space, including the Moon and other celestial bodies, States Parties to the Treaty shall be guided by the principle of cooperation and mutual assistance and shall conduct all their activities in outer space, including the Moon and other celestial bodies, with due regard to the corresponding interests of all other States Parties to the Treaty. States Parties to the Treaty shall pursue studies of outer space, including the Moon and other celestial bodies, and conduct exploration of them so as to avoid their harmful contamination and also adverse changes in the environment of the Earth resulting from the introduction of extraterrestrial matter and, where necessary, shall adopt appropriate measures for this purpose. If a State Party to the Treaty has reason to believe that an activity or experiment planned by it or its nationals in outer space, including the Moon and other celestial bodies, would cause potentially harmful interference with activities of other States Parties in the peaceful exploration and use of outer space, including the Moon and other celestial bodies, it shall undertake appropriate international consultations before proceeding with any such activity or experiment. A State Party to the Treaty which has reason to believe that an activity or experiment planned by another State Party in outer space, including the Moon and other celestial bodies, would cause potentially harmful interference with activities in the peaceful exploration and use of outer space, including the Moon and other celestial bodies, may request consultation concerning the activity or experiment.

The following analysis naturally echoes the most important of these approaches, but also extends to more recent views that have emerged in legal scholarship. As will be shown, these views either reinforce some of the existing arguments or introduce new ones. The subsequent theoretical analysis is arranged along thematic lines corresponding to the arguments advanced.

3.1 The Issue of the Utilization and Exploitation of Space Resources Should Be Addressed in Conjunction with Other Constraints and Activities in Outer Space

In theory, it has been argued that an effective consideration or regulation of the issue of space resources should be made, *inter alia*, in conjunction with other space activities, which are somehow related to the issue in question and must necessarily be taken into account in order for the consideration or regulation sought to be optimal. The most important of these interconnections are presented below.

3.1.1 *The Duty to Avoid "Forward" and "Back Contamination"*

The need to consider the environmental impacts of the use and exploitation of space resources has been proposed on several occasions in theory. This is because, as has been rightfully pointed out, "…the present projects focused on the extraction and use of space resources envisage the transport of a space object from the Earth to outer space, its impact with a celestial body, and the use of space resources either in outer space or their return to the Earth. In the more distant future, the launching of space objects from the Earth can be substituted by building the entire space structures in outer space. This opens the question on the applicability of Art. IX OST interpreted by COSPAR rules to the space mining activities".[1] The legal basis for environmental protection in the context of space activities in general is provided by Article IX OST.

According to this provision, the so-called "forward contamination", i.e. harmful contamination of celestial bodies, must be avoided. The same, however, applies to "backward contamination", which can occur if extraterrestrial matter is introduced into the Earth's environment. It is obvious that both situations can, under certain conditions, occur in the context of the use and exploitation of resources originating from celestial bodies. Hence, concerns have been expressed in relation to the damage that space mining operations could cause in the environment of outer space and around the Earth, which would constitute a breach of the environmental obligations prescribed in Article IX of the Outer Space Treaty. Dr. Gbenga Oduntan was of the view that the multiplication of space mining operations could lead to an "irrevocable alteration" of "the pristine conditions of the cradle of nature from which our own Earth was born […]—making it harder to trace how we came into being"; besides, forward contamination of celestial bodies due to such human activities could destroy our chances of discovering alien life.[2]

In this context, it has been argued that the COSPAR (International Committee on Space Research) recommendations on the contamination of celestial bodies[3] should, under certain conditions, also apply to "space mining".[4] COSPAR recommendations

---

[1] M. Hofmann, "COSPAR Recommendations in a New Context: Environmental Aspects of Space Mining", Proceedings of the International Institute of Space Law, 59, 2016, p. 210. See also the general reflection *in:* G. Leterre, "Space Mining and Environmental Protection: Recycling International Agreements into New Legal Practices", Proceedings of the International Institute of Space Law, 62, 2019, p. 83–94.

[2] See K. Pascual, "U.S. Space Mining Law Is Potentially Dangerous And Illegal: How Asteroid Mining Act May Violate International Treaty" (28 November 2015) online:
<http://www.techtimes.com/articles/111534/20151128/u-s-space-mining-law-is-potentially-dangerous-and-illegal-how-asteroid-mining-act-may-violate-international-treaty.htm> (Last accessed: 13.2.2024).

[3] COSPAR Policy on Planetary Protection, see https://cosparhq.cnes.fr/assets/uploads/2020/07/PPPolicyJune-2020_Final_Web.pdf (Last accessed: 13.2.2024).

[4] M. Hofmann, "COSPAR Recommendations…", *op. cit.*, p. 210.

are mainly related to information sharing, so this process can be important for and can also affect the governance of resources of celestial bodies, both to avoid "forward contamination" and "backward contamination".[5]

It has, however, also been argued that Article IX OST, insofar as it refers to regulations of an environmental nature, should be interpreted from an anthropocentric perspective: "an anthropocentric approach would bring Article IX in line with the Outer Space Treaty as a whole. The purpose of the Outer Space Treaty, as well as of the rest of the *corpus juris spatialis*, is not the preservation of the space environment in its pristine state, but rather the exploration and use of outer space for peaceful purposes".[6]

A different approach was expressed by the author of this book, after having noticed the increased anthropocentrism of the Outer Space Treaty (mainly due to the concept of 'province of mankind' and the content of Article I OST), by considering that Article IX OST was moving in the opposite direction:[7]

> This anthropocentrism is somewhat tempered by Article IX of the OST. Said article is the only hard law provision to rely upon, as it sets some standards in order to mitigate contamination of the celestial bodies from human missions (*forward contamination*) or contamination of the Earth by the introduction of extra-terrestrial matter (*backward contamination*). Thus, *Article IX is the only legal provision so far that moves, to a certain extent, away from the anthropocentric logic of the space law in force*, as it could serve as the legal "basis for the environmental protection of outer space" from human intervention.[8] Furthermore, Article 7 par. 1 of the Moon Agreement constitutes an advanced version of the same principle, as it gives priority to the forward protection of the Lunar environment, by stating that "In exploring and using the Moon, States Parties shall take measures to prevent the disruption of the existing balance of its environment, whether by introducing adverse changes in that environment, by its harmful contamination through the introduction of extra-environmental matter or otherwise".[9] In this context, the *Galileo* case should be mentioned: said spacecraft plunged into Jupiter's atmosphere on 21 September 2003, thus being deliberately destroyed in order to protect a possible ocean beneath the icy crust of the moon Europa (of Jupiter)[10] *(emphasis in the original).*
>
> As such, and in the absence of sufficient positive law provisions which would ensure the safeguard of extra-terrestrial natural ecosystems and the protection of existing life forms, intelligent or not – with the exception, of course, of Article IX of the OST -, a Universe –

[5] *Idem.*

[6] F. Bergamasco, "Space Mining and the Protection of Extra-Terrestrial Environment in the Light of Article IX of the Outer Space Treaty", Proceedings of the International Institute of Space Law, 60, 2017, p. 169, citing C. Cinelli & K. Pogorzelska, "The Current International Legal Setting for the Protection of the Outer Space Environment: the Precautionary Principle Avant La Lettre", Reciel 22 (2), 2013. See also *infra*, under V.1.2.2.3.

[7] George D. Kyriakopoulos, "Where Law Meets Cinema: James Cameron's Avatar as Food for Thought About the Anthropocentric Nature of Space Law", Proceedings of the international Institute of Space Law, 58, 2015, p. 310.

[8] S. Marchisio, "Article IX", *in:* S. Hobe, B. Schmidt-Tedd, K.-U. Schrogl & G. Meishan Goh (eds.), Cologne Commentary on Space Law, Vol. 1 Outer Space Treaty, Carl Heymanns Verlag, 2009, p. 176.

[9] See Marchisio, "Article IX", *op. cit.*, p. 177.

[10] http://solarsystem.nasa.gov/galileo/; Marchisio, *op. cit.*, p. 179.

> "province of Mankind" is a legal construction that could give rise to a version of space exploitation similar to the one of the colonial past of the Earth.[11]

In any case, the obligation to prevent "forward" and "back" contamination, as introduced by Article IX of the Outer Space Treaty and further elaborated in the COSPAR recommendations, cannot be ignored in the future planning and implementation of the exploitation of space resources, under whatever regime it may ultimately take place.[12] Even more so as the need for a sustainable use of outer space, which is now universally accepted, is becoming undeniable. Although the need for sustainable use of space in general is not exactly a theoretical consideration, but a political option at the international level (mainly through United Nations action), its direct relevance to environmental issues in space (e.g. in the context of planetary protection) requires that it be addressed here. This is supported by the obvious importance of the issue of space resources for the advancement of sustainability in space, as will be shown below.

3.1.2 The Need for Sustainable Space Activities

Sustainability in the context of space activities is understood within the broader debate, in the United Nations, on the 2030 Agenda for Sustainable Development,[13] and the promotion of the 17 Sustainable Development Goals (SDGs).[14] Space sustainability is driven by the need to ensure that the current use of space—mainly through the activities of the so-called space-faring States—does not compromise its future use, by all users and, in particular, by developing countries. General Assembly Resolution 76/3 (2021) referred to space "as a driver of sustainable development"

---

[11] Cf. Bueckling observing that "the very notion of heritage, taken in relation to the concept of mankind, marks the birth of an ancient human norm" – A. Bueckling, "The Strategy of Semantics and the 'Mankind Provisions' of the Space Treaty", J.S.L., vol. 7 no 1, 1979, p. 17, at 21. For the same author, the concept of 'Mankind', for the time being, "does not represent a workable legal term"—*op. cit.*, p. 22.

[12] However, the extent to which the international community is genuinely committed to preventing forward contamination remains to be verified in practice—see also *infra*, under V.1.2.2.

[13] For the concept of sustainable development in international law, see V. Barral, "Sustainable Development in International Law: Nature and Operation of an Evolutive Legal Norm", E.J.I.L, Vol. 23 no. 2, 2012, p. 377 *et seq.*; C. Voigt, Sustainable Development as a Principle of International Law: Resolving Conflicts between Climate Measures and WTO Law, NIjhoff, 2009, p. 11 *et seq.*; N. Schrijver & F. Weiss (eds.), International Law and Sustainable Development: Principles and Practice, Nijhoff, 2004, p. 7 *et seq.*

[14] In his 2023 Policy Brief 7, the Secretary-General highlighted the "important link between outer space and the 2030 Agenda for Sustainable Development", noting that "today, nearly 40% of the targets underpinning the Sustainable Development Goals leverage Earth observation and global navigation satellite systems"—Secretary-General Policy Brief on Outer Space: Our Common Agenda Policy Brief 7 For All Humanity—the Future of Outer Space Governance (May 2023), A/77/CRP.1/Add.6, p. 11, https://www.unoosa.org/oosa/en/documents/doc/policy-brief.html (last accessed: 13.02.2024) (hereinafter: "Policy Brief 7").

and defined, in the context of the so-called "Space2030" Agenda, "four overarching objectives ... structured around the four pillars of space economy, space society, space accessibility and space diplomacy". The first among these objectives is to "enhance the economic benefits derived from space and strengthen the role of the space sector as a major driver of sustainable development", whereas the third one refers to the need to "improve access to space for all and ensure that all countries can benefit socioeconomically from space science and technology applications and space-based data, information and products, thereby supporting the achievement of the Sustainable Development Goals". Finally, the fourth "overarching objective" clearly promotes multilateralism in the context of space activities, as it explicitly refers to "Build[ing] partnerships and strengthen[ing] international cooperation in the peaceful uses of outer space and in the global governance of outer space activities".[15] The subsequent Resolution 77/121 also emphasizes the need for international cooperation, stating that:

> …regional and interregional cooperation in the field of space activities is essential to strengthen the peaceful uses of outer space, assist Member States in the development of their space capabilities and contribute to the implementation of the 2030 Agenda for Sustainable Development….[16]

It is no coincidence that in his "Policy Brief 7" of 2023, the Secretary-General recommends that "the Committee on the Peaceful Uses of Outer Space could consider developing new governance frameworks for various areas of space sustainability. Such frameworks, … would comprise separate, mutually reinforcing instruments…". Space resources are explicitly mentioned as one of the issues to be addressed:

> Space resource activities. Develop an effective framework for sustainable exploration, exploitation and utilization of the Moon and other celestial bodies. This framework could include binding and non-legally binding aspects and should build upon the five United Nations treaties on outer space and other instruments for international cooperation in the peaceful uses of outer space.[17]

From the above, a crucial parameter for the future exploration and use of space resources is introduced: the need to do so in a sustainable manner. According to the above-mentioned "overarching objectives", such sustainable exploration and exploitation of space resources will be achieved through international cooperation and should ensure that all countries benefit in some way, so that space will indeed become "a major driver of sustainable development". In today's overall reflection on the future regulation of space resources, there is an important requirement which, it should be noted, is not disputed by the international community (however it may be understood and interpreted): the need for sustainable use and exploitation of space resources.

[15] UNGA Resolution 76/3, "The 'Space2030' Agenda: space as a driver of sustainable development", 25 October 2021, A/RES/76/3.

[16] UNGA Resolution 77/121, "International cooperation in the peaceful uses of outer space", 12 December 2022, A/RES/77/121.

[17] Policy Brief 7, p. 19.

3.1.3 *Exploitation of Space* vs. *Space Exploration*

From the beginning of the Space Age, one of the fundamental principles of the law regulating space activities was the freedom of exploration and use of outer space. Whereas "use" refers to the utilization/exploitation of outer space for economic and/or non-economic ends,[18] "exploration" is "the activity of searching and finding out about something", but also "the act of traveling to a place or searching a place in order to learn about it".[19] The latter definition, focusing on the concept of travel, better reflects the sense of space exploration.[20]

Whether exploration and exploitation are seen as conflicting or synergistic concepts, they are two sides of the same coin. In the broader context of a systemic analysis, these concepts describe the ways in which a system adapts to its environment. Exploration has been defined as "the pursuit of what might come to be known, involving search, variation, risk taking, experimentation, play, flexibility and innovation"—while exploitation is "the refinement and implementation of what is already known, involving choice, selection, efficiency, standardization, reliability and control".[21] The same analysis points to the following relationship between the two concepts:

> While exploration without exploitation leads to a cascade of experiments without the development of competence in any of them or discrimination among them, exploitation without exploration leads to stagnation and failure to discover new, useful directions... Therefore, maintaining an appropriate balance between exploration and exploitation is a primary factor in a system's survival and prosperity in the face of changing environments...[22]

Systematically applying the above findings to space—assuming it is regarded as an 'environment' of the Earth system—it follows that while space exploration has as its mission the transformation of the unknown (space environment) into the 'known', exploitation in the same space will have to do with the selection, control and general use of that (part of the space environment) which has already become known. Consequently, space exploration without subsequent exploitation/use of space will be a 'sterile' activity, whereas (space) exploitation simply cannot take place without prior exploration.

The Outer Space Treaty explicitly guarantees both the freedom to explore and the freedom to use outer space. Article I, para. 2 refers to the freedom of "exploration and use by all States without discrimination of any kind, on the basis of equality

[18] See S. Hobe, "Article I", *in:* S. Hobe, B. Schmidt-Tedd, K.-U. Schrogl & G. Meishan Goh (Eds.), Cologne Commentary on Space Law, Vol. 1 Outer Space Treaty, Carl Heymanns Verlag, 2009, p. 21 (hereinafter "CoCoSL 1").

[19] https://dictionary.cambridge.org/dictionary/english/exploration (last accessed: 14.02.2024).

[20] See G.D. Kyriakopoulos, "Colonies on the Moon (and/or Mars)? New Challenges for International and National Law", Proceedings of the International Institute of Space Law, 2020, p. 170.

[21] M. Augier & D.J. Teece (eds.), The Palgrave Encyclopedia of Strategic Management, Palgrave MacMillan, 2018, p. 543.

[22] *Idem.*

and in accordance with international law", while para. 3 of the same article establishes the freedom of scientific investigation in outer space. Scientific research is part of the broader exploration,[23] while "use" as a concept clearly includes the economic exploitation of outer space.[24] Obviously, the principle of non-appropriation (Article II OST) in some way limits the freedom of use of Article I para. 2, but this question is not at issue here.

The author had pointed out in a previous contribution[25] that space exploration has essentially come to a standstill compared to what happened in the 50s, 60s and 70s,[26] and further noted that a substantial resumption of space exploration is a factor that can have a significant impact on the issue of space resource exploitation:

> It seems that at present the international community is mainly concerned with the challenge of initiating activities of space *exploitation* – in other words with the *stricto sensu* use of outer space and not with space exploration. Several important legal issues are associated with this new reality. The thorniest question in this regard is whether positive space law can adequately regulate the strongly emerging private activities aimed at the economic exploitation of celestial bodies.[27] Space law has been expressed so far in broad, vague principles that have permitted the maximum flexibility necessary for exploratory space activities.[28]

---

[23] "Exploration is the general finding out of something yet to be explored, which may or may not include scientific activity"—Hobe, CoCoSL 1, p. 34.

[24] "…'use' as the broader and more general term certainly includes exploitation of outer space and of celestial bodies"—Hobe, CoCoSL1, p. 35.

[25] G.D. Kyriakopoulos, "Where Law Meets Cinema: James Cameron's Avatar as Food for Thought About the Anthropocentric Nature of Space Law", in Proceedings of the international Institute of Space Law, vol. 58, 2015, p. 308.

[26] On this issue, see J. Arnould, "Chapter 11: An Urgent Need to Explore Space", *in:* J.S.J. Schwartz & T. Milligan (eds.), The Ethics of Space Exploration, Springer, 2016, p. 153 *et seq.*

[27] See, in particular: R.J. Lee, Law and Regulation of Commercial Mining of Minerals in Outer Space, Springer, 2012, 372 p.; Z. Meyer, "Private Commercialization of Space in an International Regime: A Proposal for a Space District", Northwestern J.I.L.&B., Vol. 30 Issue 1, 2010, p. 241–261; S. Hobe, "The Impact of New Developments on International Space Law (New Actors, Commercialization, Privatization, Increase in the Number of "Space-faring Nations"), Uniform Law Review, 2010, Vol. 15, issue 3–4, p. 869–881; F. Tronchetti, The Exploitation of National Resources of the Moon and Other Celestial Bodies, Nijhoff, 2009, 320 p.; J.L. Zell, "Putting a Mine on the Moon: Creating an International Authority to Regulate Mining Rights in Outer Space", Minnesota J.I.L., vol. 15, 2006, p. 489–519; R. Jakhu, "Twenty Years of the Moon Agreement: Space Law Challenges for Returning to the Moon", Z.L.W., vol. 54, 2005, p. 243–260; W. White, "The Legal Regime for Private Activities in Outer Space", paper presented at "Space: The Free Market Frontier", 15 March 2001, *in:* http://www.spacefuture.com/archive/the_legal_regime_for_private_activities_in_outer_space.shtml (last accessed: 16.04.2023); K.M. Zullo, "The Need to Clarify the Status of Property Rights in International Space Law", Georgetown L.J., vol. 90. 2001–2002, p. 2413–2444.

[28] As Bueckling rightly observed, "time and again it becomes apparent how difficult it is to provide adequately phrased rules for, and to systematize in legal language the extremely complicated subject matter created by the technological explorations in outer space and the resulting multitude of conflicting interest"—Adrian Bueckling, "The Strategy of Semantics and the 'Mankind Provisions' of the Space Treaty", J.S.L., vol. 7 no 1, 1979, p. 17. For the same issue, see Heidi Keefe, "Making the Final Frontier Feasible: A Critical Look at the Current Body of Outer Space Law", Santa Clara High Technology Law Journal, Vol. 11 Issue 2, 1995, p. 346.

> Nevertheless, law, in general, is not immutable but responds to the needs of society. As exploration has given way to exploitation and as this - predominantly international - law lacks the specificity and legal certainty necessary for mature commercial activity, sooner or later law will follow the international community needs in outer space.
>
> *Anyway, every systematic effort of exploitation of natural resources in outer space will likewise sooner or later require a corresponding promotion of space exploration. Otherwise, the scope for commercial exploitation will remain narrow, essentially limited to a few sub-orbital recreational activities as well as mining activities on near-Earth celestial objects (NEOs).* Add to this the parameter of Earth's overpopulation with its environmental implications and the need to seriously tackle the prospect of space exploration follows effortlessly. All roads lead to the Stars, regardless of the existing difficulties (*Per aspera ad astra*)[29] (emphasis added).

This position—that the exploitation of space cannot take place in isolation from its exploration, which must be continuous—and that, in particular, the issue of space resources must be addressed, both technically and legally, in conjunction with space exploration—seems to be accepted by NASA, as it is mentioned in one of its websites:

> The reasons to explore the universe are as vast and varied as the reasons to explore the forests, the mountains, or the sea. Since the dawn of humanity, people have explored to learn about the world around them, *find new resources*, and improve their existence (emphasis added).[30]

In conclusion, the importance of space exploration for the future use of space resources is most evident at the present time, as the human species appears to be preparing to set foot on celestial bodies for the first time since 1972.[31] Through Space Policy Directive 1 (2017), the United States have already expressed their intention to "...lead the return of humans to the Moon for long-term exploration and utilization, followed by human missions to Mars and other destinations".[32] Russia and China, for their part, announced, in 2018, that they plan to jointly build a lunar base.[33] Since 2015, the European Space Agency (ESA) has put forward the multinational Moon Village concept, which aims to be "a collaboratively designed and expandable permanent lunar settlement".[34] Finally, on Mars 2021, three different missions arrived at Mars: The UAE space probe Hope, the Chinese Tianwen-1

[29] Kurt Vonnegut, The Sirens of Titan, 1959. See, again, Kyriakopoulos, "When Law.", *op. cit.*

[30] NASA, "Why Go to Space?", https://www.nasa.gov/humans-in-space/why-go-to-space/#.VqU3nzaAXOg (last accessed: 17.02.2024).

[31] See C.D. Johnson, "Renewed Ambitions in Space-Exploration Lawmaking", Air and Space Lawyer, vol. 33, no. 3, 2020, p. 20.

[32] https://2017-2021.state.gov/space-policy-directive-1-reinvigorating-americas-human-space-exploration-program/ (last accessed: 17.2.2024).

[33] See https://www.cnbc.com/2020/07/15/russia-space-chief-dmitry-rogozin-dismisses-nasas-moon-program-considering-china-lunar-base.html (last accessed: 17.02.2024). For the International Lunar Research Station, see IV.3.

[34] https://www.astronomy.com/observing/moon-village-humanitys-first-step-toward-a-lunar-colony/ (last accessed: 17.02.2024).

orbiter and NASA's Perseverance rover.[35] It is clear that space exploration is entering a new and very interesting phase, which will inevitably affect the emerging international legal framework for the use of space resources.

3.2 Space Resources Must Be Exploited Through an International Legal Regime

The precise meaning of the term 'common heritage of mankind', which is used to describe the status of the Moon and its natural resources in Article 11 of the 1979 Moon Agreement, has also been sought in theory. In this context, it has been argued that said term essentially refers to the fact that "the legal regime governing outer space resources must be international and not national".[36]

In this context, this opinion essentially suggests the need for an *international* legal regime for the exploitation of space resources, modelled on Article 11 para. 5 of the 1979 Moon Agreement. This regime is opposed to unilateral legislative initiatives by some States, such as the 2015 U.S. "Commercial Space Launch Competitiveness Act". Further, a review is made of the proceedings of the 55th session of the Legal Subcommittee of the UNCOPUOS, which show that, at that time, the States that were parties to the Moon Agreement, by and large, maintained a wait-and-see attitude towards the then-newly introduced U.S. Act. The position of Belgium is, however, particularly highlighted, which, in essence, strongly advocated the need for an international legal regime for space mining:

> Outer space natural resources cannot be appropriated through extension of national jurisdictions. Such appropriation, besides its incompatibility with international law, would mean risk of global economic imbalance because of exclusivity claimed by some to the detriment of others. Even if such an imbalance has not been envisaged in the short or medium term, any precedent which will be derived from any national claim over such resources could determine virtually irrevocably an advantageous position or a position of disadvantage of states regarding these resources and their derived products.[37]

In the same pattern, the view is expressed that Article II OST, containing the principle of non-appropriation, is also applicable to the acts of private individuals, since their activities are also considered 'national' activities, in accordance with the wording of Article VI OST. The invocation of the non-appropriation principle as a fundamental principle of space law, which constitutes a legal barrier to the establishment of property rights in favor of individuals under national laws (and,

---

[35] See the relevant report in: https://www.planetary.org/articles/mars2021 (last accessed: 17.02.2024).

[36] I. Marboe, "The End of the Concept of Common Heritage of Mankind", Proceedings of the International Institute of Space Law, 59, 2016, pp. 225–238.

[37] Marboe, *op. cit.*

consequently, on a unilateral basis), is perhaps the strongest argument of those who deny the possibility of national regulation of the governance of space resources.[38] The conclusion, also according to this reasoning, is the necessity of a "detailed and internationally accepted legal regime, as rightfully mentioned in Article 11 of the 1979 Moon Agreement, in order to ensure a balance of interests of the private sector eager to exploit new opportunities and gain profit, and the government bound by its international obligations, including full responsibility for national space activities".[39]

3.3 To Ensure Optimal Use and Exploitation of Space Resources, the Outer Space Treaty Must Be Amended

In space law theory, the view is also expressed that the Outer Space Treaty, as it exists today, is not sufficient to effectively regulate the use and exploitation of space resources, and therefore this treaty should be amended. This view is part of a broader consideration that the OST should be revised because, in general, it does not meet modern space needs.[40] With respect to space resources, the difference of opinion is to do with the direction such an amendment should take. According to an academic approach, the OST should be amended towards the adoption of a model benefit-sharing agreement, between private entities and States wishing to operate in the field of space resources. This agreement would confer "benefits to the private companies analogous to their investment, as well as to the authorizing States the right to collect and further distribute benefits deriving from the undertaken activities".[41]

In the same context of amending the Outer Space Treaty, the abrogation of Article II and the principle of non-appropriation contained therein has been advocated. However, such a possibility has been rightly described as "counterproductive to the interests of space commercialization", while it would lead to a "'space rush' with a clean slate of celestial treasures open and available to be grabbed by the

[38] See O. Volynskaya, "Space Resources Exploitation from the international and Domestic Law Perspectives: The Russian Approach", Proceedings of the International Institute of Space Law, Vol. 59, 2016, p. 293 *et seq*.

[39] Volynskaya, *op. cit.*

[40] R. Schembri, "The Outer Space Treaty Needs Amending", Human Space Program, January 25, 2023, https://medium.com/@humanspaceprogram/the-outer-space-treaty-needs-amending-d9813c40ba1b.

[41] M. Manoli, "Pondering the Legitimacy of the Outer Space Treaty from the Perspective of Space Natural Resources Exploitation: Is It Time to Act instead of React?", Proceedings of the International Institute of Space Law, Vol. 60, 2017, p. 268 *et seq*.

quickest or the strongest". It would also allow for the emergence of growing and often overlapping claims by space-faring States over celestial bodies.[42]

In addition, other voices have called for an amendment to Article II OST and the "province of mankind" principle to facilitate the use of space resources in the context of a future Mars settlement:

> …Could we not simply extend our stay and increase our harvest of materials and remain within the bounds of the treaty? This is an important question to ask, and one that is not clearly answered in the wording of the OST. … By claiming only the resources in a radius of land and allowing sovereignty, or control of rules and people, only within the terrestrial space stations themselves, no country will have sovereign reign over Mars territory. This method will have its own challenges as colonies develop….[43]

Regarding the amendment of the Outer Space Treaty, Article XV provides that:

> Any State Party to the Treaty may propose amendments to this Treaty. Amendments shall enter into force for each State Party to the Treaty accepting the amendments upon their acceptance by a majority of the States Parties to the Treaty and thereafter for each remaining State Party to the Treaty on the date of acceptance by it.

Therefore, the Convention does not provide for a collective revision mechanism but leaves this prospect to the initiative of the Member States. This does not, however, mean that a collective movement for such a revision cannot be excluded under the corresponding articles of the 1969 Vienna Convention on the Law of Treaties (Articles 39–41). In any case, for such a revision to have global significance and effect, the General rule regarding the amendment of treaties in Article 39 VCLT, namely that "A treaty may be amended by agreement between the parties", must be fulfilled. This is because any amendment to the OST to effectively address space resource issues would have to be universally accepted to be practical. Therefore, the situation described in Article 40 para. 4 VCLT[44] should be avoided, i.e. that there may be States which, by not consenting to the amendment, will not be bound by it.[45] Such a possibility is also contained in Article XV OST.

Be that as it may, there is no sign of any will, even on the part of some States, to propose a revision of the Outer Space Treaty in order to legally regulate the issue of space resources. Coupled with the fact that no revision of this treaty, on any subject, has been brought about so far, this prospect is not very likely.

---

[42] See L.I. Tennen, "Second Commentary on Emerging System of Property Rights in Outer Space", Proceedings of the United Nations/Republic of Korea Workshop on Space Law "United Nations treaties on outer space: actions at the national level", Daejeon, Republic of Korea, 3–6 November 2003, 2004, ST/SPACE/22, p. 344–345. Also, by the same author, "New Developments and the Legal Framework Covering the Exploitation of the Resources of the Moon", Proceedings on the Law of Outer Space, Vol. 47, 2004, p. 522 *et seq.*

[43] S. Bruhns & J. Haqq-Misra, "A pragmatic approach to sovereignty on Mars", Space Policy, Vol. xxx, 2016, p. 6.

[44] "The amending agreement does not bind any State already a party to the treaty which does not become a party to the amending agreement…".

[45] See also, in this respect, I. Marboe, "Reviewing the Moon Agreement or Amending the Outer Space Treaty? - Views of UNCOPUOS Member States", Proceedings of the International Institute of Space Law, Vol. 62, 2019, p. 404.

3.4 Article I OST Introduces a Specific Obligation, Under Which the Legality of the Utilization and Exploitation of Space Resources Should Be Assessed

In legal theory, the question of the meaning and effect of Article I para. 1 OST with regard to the permissibility of the use and exploitation of space resources has been thoroughly discussed. This paragraph provides, inter *alia*, that "the exploration and use of outer space, including the Moon and other celestial bodies, shall be carried out for the benefit and in the interests of all countries". A requirement is therefore introduced, an orientation ("for the benefit and in the interests of all countries"), which delineates the exploration and use of outer space. To what extent does this have a real impact on the question of space resource activities? Here is an interesting, related analysis from an earlier monograph as to whether this wording introduces a substantive obligation, for Member States in the OST, or should be seen as a moral imperative only:

> The idea of Article I in practice as being no more than a moral obligation, instead of a legal one, is a view that is shared with Gorove by other commentators... Although this formulation may be considered the most favourable, especially in the context of promoting private and commercial space activities, *it must be noted that there are two indicia to suggest that the requirement actually imposes a positive duty.*[46] The first is that the requirement in Article I utilises the plural form "interests" instead of the singular, which may indicate that this involves more than "just the vague, general 'interest' of all countries" and, instead, represents specific and identifiable interests. This may be taken to mean that a particular set of interests of all States is to be taken into account in the conduct of space activities. The second is that while Article I may be considered to be "an aspiration couched in very general terms which could not be specifically implemented without further elaborations and guidelines", the provisions of the Agreement on the Activities of States on the Moon and other Celestial Bodies (the "Moon Agreement") may arguably constitute the further elaborations and guidelines to give effect to the "interests and benefits of all countries" requirement. Accordingly, even though the Moon Agreement has not won widespread acceptance as the means of implementing the requirement, this may not of itself prejudice the view that the requirement may nevertheless require implementation at a practical level.
>
> The foregoing analysis may be condensed to produce at least three possible outcomes and the corresponding applications on the exploration and extraction segments of a commercial space mining operation:
>
> (1) *Generalised mission statement rather than positive and specific duty.*[47] If the requirement of "benefits and interests of all countries" is to be regarded as a generalised mission statement for all space activities instead of the imposition of a positive and specific duty, then clearly commercial mineral exploration and extraction activities on celestial bodies may be considered a positive development for all States, notwithstanding the absence of any sharing of financial or tangible benefits to other States.

[46] Emphasis added.

[47] Emphasis in the original.

(2) *Obligation imposed on the activity rather than the results derived thereof.*[48] If the requirement does impose a specific and positive duty but such a duty is imposed on the activity itself instead of on the results and outcomes derived thereof, then the duty may be interpreted as no more than a negative duty of ensuring that the activity does not cause a detriment to any State. In such a case, commercial mineral exploration and extraction activities would not have much difficulty fulfilling such an obligation, as these activities or the means involved are unlikely to cause detriments for other States.

(3) *Positive duty to share the benefits derived from space activities.*[49] If the requirement under Article I is to be interpreted as an actual obligation to share the resulting benefits derived from space activities, whether financial, tangible or both, then the Moon Agreement is an example, though an unacceptable one, of the practical means of fulfilling this obligation. *However, it follows then that the obligation does not arise until the State or its private entities have gained a benefit that is capable of being shared on an equitable basis.*[50] In the context of a commercial space mining venture, such a benefit would be produced only in the exploitation segment and thus the obligation would have no application on the exploration and extraction segments of the venture. Further, there is no suggestion that all States would be entitled to an equal share of such benefits. This is supported, for example, by the express stipulation in the Moon Agreement that there is to be an "equitable" sharing in the "benefits" derived from mineral resources extracted from celestial bodies, rather than an "equal" sharing in the "materials" or "profits" derived from such activities.

It can be seen from above that some doubts remain concerning the legal content and effect of the requirement that space activities be carried out "for the benefit and in the interest of all countries" and, correspondingly, the legality conducting commercial mineralogical prospecting and exploitation activities on celestial bodies. The existing body of state practice would tend to suggest that the provision is at least a generalised mission statement at the very least and at most an obligation on the nature of the activity rather than the benefits derived therefrom. *However, the exploitation of mineral resources from celestial bodies may well be an activity that may attract an interpretation that involves a positive duty to share the derived benefits with other countries. This position is further supported by the existing formulation of the common heritage of mankind principle as contained in Article 11 of the Moon Agreement.*[51] Therefore, it is in the context of Article I and other relevant provisions of the Outer Space Treaty that the effect of the common heritage of mankind principle must be considered.[52]

---

[48] Emphasis in the original.

[49] Emphasis in the original.

[50] Emphasis added.

[51] Emphasis added.

[52] R.J. Lee, Law and Regulation of Commercial Mining of Minerals in Outer Space, Springer, 2012, p. 157 *et seq.*

Interestingly, the above analysis, in the context of an exhaustive hypothesis, identifies, in the context of Article I OST, the existence of a legal obligation to share the benefits of space mining activities. According to this reasoning, the duty in question is established regardless of whether the relevant provision imposes an explicit obligation to share the benefits arising from the exploration and use of outer space or merely constitutes a general recommendation (as accepted by the present author).[53] Notably, and in line with the above analysis, this positive obligation is conceptually framed within the notion of the "common heritage of mankind" as articulated in Article 11 of the Moon Agreement. Of course, if Article I para. 1 OST is not interpreted in conjunction with Article 11 MOON—particularly in light of the Moon Agreement's limited acceptance by States—it may be reasonably argued that the ambiguity surrounding the precise meaning of Article I para. 1 OST resurfaces as a legal issue. Yet, even in such a case, it appears that Article I OST, considered on its own, does not explicitly establish a right, in particular of non-governmental entities/private individuals, to appropriate space mineral resources. The same analysis also concludes that the obligation of States to operate in outer space in a manner that does not interfere with the rights of other States is underscored by the comparative application of Articles I and IX OST. This is another aspect which, for obvious reasons, is of relevance to the legal debate on space resources:

> ...the corresponding interests of States in the context of Article IX are the freedoms provided for in Article I, namely the freedom of exploration and use of outer space and access to all areas of the Moon and other celestial bodies. Consequently, it may be argued that Article I, taking into consideration Article IX, does no more than to impose a negative duty on States in their conduct of activities in outer space and on celestial bodies. This duty is to ensure that such activities do not interfere with the rights provided under the provisions of the Outer Space Treaty or cause any detriment to other States. It has been observed that the existing state practice already reflects this interpretation of the requirement under Article I. ...this conclusion poses as one of the legal obstacles to the extraction segment of a commercial mining venture in space.[54]

In any case, the freedoms of Article I OST are substantially limited by the principle of non-appropriation enshrined in Article II OST, which is discussed below.

3.5 An Alternative Interpretation of the Non-appropriation Principle (Article II OST) Is Required

Since the main legal obstacle to the legality of the exploitation of space resources is the non-appropriation principle of Article II OST, it has been suggested that it should, if possible, be sought whether this principle can be interpreted in a way that would make "space mining" legally possible. It is worth noting that, according to Article II OST, "Outer space, including the Moon and other celestial bodies, is not

---

[53] See V.1.2.3.2.

[54] Lee, *op. cit.*, p. 160.

subject. to national appropriation by claim of sovereignty, by means of use or occupation, or by any other means".

3.5.1 The Meaning of "Appropriation"

One of the important aspects of interpreting Article II OST is defining what is meant by "appropriation". To this end, it has been argued that it would be useful to look for the concept of appropriation in the context of its origins, i.e. in Roman law.[55] In view of this necessity, the concept of *res communis omnium*, as it existed under Roman law, is approached in order to verify that, as now, "*res communes omnium* were not owned by anybody and were also incapable to have a dominus, being forever considered 'out of the patrimony' of men".[56] Furthermore, a distinction is made, as regards Roman *res communes omnium*, between 'container', which remained firmly out of possession, and 'content', which had the character of a stray object and could therefore be seized:

> All things 'common' to everybody were structured as a complex category, made of two things in one. The first one – the 'container' - was the physical domain at large: the air, the flowing waters, the seas and, as a consequence, the seashores. The second one – 'the content' – was the set of all things that could be found in that domain, such as birds in the air, fish in the sea or pebbles on the seashore. Thus, if the container was regulated by the principles described above, the content was, on the other hand, considered a *res nullius*, out of anyone's patrimony (in its natural state), but susceptible to be appropriated once seized.[57]

But this is precisely where this approach reaches its limits, because the resources of the celestial bodies cannot be considered as 'content' and, consequently, susceptible to seizure:

> A different option is to say that resources are indeed exploitable, being something different (by extraction) from the celestial body itself. *This unlocks the possibility to utilize space resources, but what happens when this 'use' entails the full disruption of the resource?* Consider the example of a small asteroid that is mined until it is completely consumed. In this case, the utilization of the celestial body's resources is, in practice, equal to the appropriation of the celestial body itself. Therefore, such activity shall be considered a breach of Article II, OST. Since the appropriation of celestial bodies is not allowed, permitting the appropriation of the resources extracted, would only shift the problem: when is a celestial body disrupted by man's intervention*? Where to draw the line? There is no answer to this question, because when all bodies beyond Earth's atmosphere are legally considered celestial bodies, the utilization of space resources is either unfeasible or uncertain*[58] (emphasis added).

[55] See A. Capurso, "The Non-Appropriation Principle: A Roman Interpretation", Proceedings of the International Institute of Space Law, Vol. 61, 2018, p. 111–128.

[56] Capurso, *op. cit.*, p. 116.

[57] Capurso, *op. cit.*, p. 118.

[58] Capurso, *op. cit.*, p. 124.

This position essentially answers—in the negative—the theoretical hypothesis of whether *enterprise rights* can be established over space resources, on the basis of their mere use and not on the basis of their ownership, so that there is no question of violation of the principle of non-appropriation of Article II OST.[59] This position is qualitatively different from the assumption of the existence of *property rights* over the resources of celestial bodies,[60] corresponding to the law of property as it exists in all national laws on Earth. The thesis on the feasibility of enterprise rights over space resources is essentially based on the International Telecommunication Union's concession of orbital positions, which constitutes a use, not an appropriation, of the orbits.[61] As has been pointed. Out, "the very distribution of allotments and subsequent entry into force of the plans adopted by the ITU do not violate the non-appropriation principle".[62] The fundamental distinction, however, lies in the fact that any allocation of orbital positions around the Earth constitutes a use that does not result in the depletion of orbital resources, whereas it is difficult to conceive of any commercial exploitation of space (mineral) resources that would not lead to their eventual exhaustion.

It has also been rightly pointed out that property rights to space resources cannot be acquired by accession either. Accession is a legal institution that existed in Roman law and is found in a variety of modern national legal systems, and refers to the acquisition of property either by labor or by the addition/mixing of new materials. The working hypothesis was whether accession *via* 3D printing, using space materials, could lead to the establishment of ownership of this material. The question is ultimately answered in the negative, since:

> ...there are two conflicting views on the scope of the non-appropriation principle: according to one view the non-appropriation principle covers exploitation of space resources, while the opposing view claims that exploitation of space resources is part of the freedom to use outer space.
>
> If we follow the first view, then there is no gap in international law that could be filled by recourse to general principles of law. Even if we assumed that (a) the general principles

[59] For this hypothesis, see L.I. Tennen, "Towards a New Regime for Exploitation of Outer Space Mineral Resources", Nebraska Law Review, Vol. 88, No. 4, 2010, p. 797 *et seq*, as well as L.I. Tennen, "Enterprise Rights and the Legal Regime for Exploitation of Outer Space Resources", U. Pac. L. Rev., Vol. 47, 2017, p. 285 *et seq*. Available at: https://scholarlycommons.pacific.edu/uoplawreview/vol47/iss2/14.

[60] For support of the existence of property rights over space resources, see indicatively: J.E. Willis, "Sovereignty in Space: Finding a Source of Private Property Rights in the Final Frontier", The Elon Law Journal, Vol. 14, 2022, p. 361–385; J.W. Benson, "Space Resources: First Come First Served", Proceedings on the Law of Outer Space, Vol. 41, 1998, p. 46–49; W.N. White Jr., "Real Property Rights in Outer Space The Future Applications of the Outer Space Treaty", Proceedings on the Law of Outer Space, Vol. 40, 1997, p. 370.

[61] Tennen, "Towards...", *op. cit.*, p. 799.

[62] Ph. De Man, Exclusive Use in an Inclusive Environment - The Meaning of the Non-Appropriation Principle for Space Resource Exploitation, Springer, 2016, p. 339. He adds further that "whatever the legal nature of the rights embedded in the various allotments, they cannot be qualified as property rights, for there is not only no use of the allotment, but also no enforceable exclusion because of it" (p. 340).

> of law to be on equal footing with international treaties and customs, and (b) accession is a general principle of law, then the non-appropriation principle (Art. II OST) would prevail as *lex specialis* in space law.
>
> If we follow the second view, then there would be no need to examine whether the non-appropriation can be circumvented through sophisticated legal constructions. Nevertheless, in this case we would have a legal gap as to the establishment of property rights, which necessitates the examination of the rest of the requirements of Art. 38(1)(c) of the ICJ statute.[63]

A similar question was raised as to whether the principle of non-appropriation would be *de facto* circumvented in the case of the establishment of settlements on celestial bodies (e.g., on the Moon) with respect to space resources that would lie within the larger perimeter of the settlement. But even in this case,

> it can be held that the resources commandeered by the operating authority should not exceed the requirements of the station, and should be in line with the goal of the activities undertaken at the facility...[whereas] the status of resources in the soil or subsoil of a celestial body is distinct from that of the surface occupied by a station, and that any assertion of rights over the latter cannot affect the property rights over the resources underneath it.[64]

in the face of the apparent impasse, the first mentioned approach reiterates the (interesting) idea of a conventional differentiation between celestial bodies—limiting the relevant concept essentially to planets—and other "bodies" (mainly asteroids up to a certain diameter), which could be considered "content" and, consequently, would be susceptible to appropriation:

> ...It is possible, then, to distinguish between celestial bodies in the legal sense (*res communes omnium*) and all the other extra-atmospheric things (*res nullius*). The conventional limit can be at 550 km of diameter, which is the size - rounded up - of the biggest asteroid....[65]

Or, as another distinguished scholar has pointed out,

> Of course, it is possible that by future international treaty or other agreement, certain objects or categories of objects will be the subject of a specific instrument with special rules and regulations.[66]

This view already existed in theory and was indicative of the search for transgressions in legal thought in the face of the deadlocks implied by the inherent rigidity of the principle of non-appropriation:

> Determining what constitutes in law a celestial body is... central for the study of property rights in outer space. Pursuant to the non-appropriation principle of Article II of the OST, celestial bodies cannot be appropriated. In practice, should [some] asteroids and comets be considered celestial bodies, they would fall under this prohibition; per *a contrario*, if they are not celestial bodies, they may become the object of private property rights... asteroids

---

[63] M. Chatzipanagiotis, "3D Printing Using Material from Celestial Bodies: A Method to Circumvent the Non-Appropriation Principle", Proceedings of the International Institute of Space Law, Vol. 59, 2016, p. 247–258, at p. 255–256.

[64] De Man, Exclusive Use..., *op. cit.*, p. 389.

[65] Capurso, *op. cit.*, p. 126.

[66] L.I. Tennen, "Enterprise Rights...", *op. cit.*, p. 284.

> and comets...with the appropriate technology, they could be moved; and they can be destroyed, i.e. consumed in their totality. Thus, they may qualify as movables.[67]

It should be noted in this respect that this analysis finally confirms the inherent difficulty of overcoming the non-appropriation principle and accepts that the optimal solutions for the utilization and exploitation of space resources will emerge, at the legal level, through processes of a conventional nature.

3.5.2 *The Scope of "National Appropriation"*

The—aforementioned—earlier monograph on the subject (by R.J. Lee) also demonstrates the importance of Article II OST in identifying—or seeking—a legal framework for space resources. This research is mainly concerned with the interpretation of the term 'national appropriation' and discusses its scope, i.e. whether it also covers relevant activities of private bodies.

This monograph documents, at a first level, the disagreement among scholars as to the actual meaning of the term "national appropriation":

> ...there remains a significant degree of disagreement among commentators even on the effect of Article II on exclusive claims of title asserted by non-governmental entities, such as private individuals or companies.[68]

In the context of this academic debate, the author seems to conclude—without being entirely clear—in favour of the view that the term 'national appropriation' also covers the property rights of private individuals, with reference to Christol's position on the matter.[69]

It is particularly interesting, however, that this study expresses the view that a broad interpretation of the term "national appropriation"—including relevant private rights—has become customary through the available practice of States in this regard.[70]

Always regarding the scope of the concept of "national appropriation", another interesting line of argument has been put forward, according to which no explicit reference to the non-appropriation of space by private individuals is required, since they will in any case operate in space on the basis of an authorization granted by a State, and therefore any restrictions imposed on States will extend to private individuals:

---

[67] V. Pop, Who Owns the Moon? Extraterrestrial Aspects of Land and Mineral Resources Ownership, Springer, 2009, p. 50–51. See also Ph. De Man, "Commercial Exploitation of Outer Space and Celestial Bodies - A Functional Solution to the Natural Resource Challenge", Proceedings of the International Institute of Space Law, Vol. 53, 2010, p. 30.

[68] Lee, *op. cit.*, p. 178.

[69] Lee, *op. cit.*, p. 178–179. The reference is to C.Q. Christol, "Article 2 of the 1967 Principles Treaty Revisited" A.A.S.L., Vol. 9, 1984, p. 263.

[70] Lee, *op. cit.*, p. 170–171.

> The argument that article II of the Outer Space Treaty does not apply to private entities since they are not expressly mentioned therein has been made many times, but must fail for the simple reason that private entities do not need to be expressly listed in article II to be fully subject to the non-appropriation principle. Private entities have the right to conduct activities in space and on celestial bodies, but that right is not unlimited, and is subject to the rule of law. Significantly, the private entity must be authorized to conduct an activity in space by the appropriate state of nationality (Outer Space Treaty, article IX). But if the state is prohibited from engaging in certain conduct, then it lacks the authority to license its nationals or other entities subject to its jurisdiction to engage in that prohibited activity…
>
> If a state could license its nationals to "privately appropriate" areas of the moon and other celestial bodies, notwithstanding the prohibition against national appropriation in article II of the Outer Space Treaty, then why could a state not also authorize its nationals to conduct other activities, in their capacity as private entities, in contravention of other articles of the Treaty? What is to prevent a state, under that scenario, from licensing its nationals to place nuclear weapons or other kinds of weapons of mass destruction in Earth orbit or on celestial bodies, notwithstanding the prohibitions contained in article IV of the Outer Space Treaty? After all, private entities are not mentioned in that article. Why stop there? Why could a state not "privatize" its nuclear testing procedures, and license a private entity to conduct nuclear weapons tests above ground, in the atmosphere, or in outer space, contrary to the provisions of the Nuclear Test Ban Treaty…, which also does not mention private entities? Carried to its logical conclusion, this argument would negate every bilateral or multilateral agreement ever made, since the states party thereto could engage in every activity they agreed to restrict or limit by the convenient subterfuge of conducting the activity through the guise of the private rather than the public sector.[71]

Moreover, according to the same line of reasoning, the fact that the term "national" is defined in Article VI of the OST to include space activities of non-governmental entities has the direct consequence that the term "national appropriation" as defined in Article 2 of the OST also includes the corresponding activities of private individuals:

> Article II prohibits "national" appropriation. The term "national" is defined in Article VI of the Outer Space Treaty to include activities that are conducted by governmental as well as non-governmental entities.35 Thus, a claim of appropriation, even if made by a non-governmental entity, is nonetheless a national activity, and is thereby prohibited by Article II.[72]

The scope of Article II of the OST has also been addressed in the context of a 2004 statement by the International Institute of Space Law (IISL), which took a clear position in favour of interpreting the provision as prohibiting the recognition of property rights by private entities. Furthermore, the IISL went a step further by pointing out that States must ensure that transactions involving the granting of such

[71] P.M. Sterns & L.I. Tennen, "Privatering and Profitering on the Moon and Other Celestial Bodies: Debunking the Myth of Property Rights in Space", Adv. Space Res., Vol. 31, No 11, 2003, p. 2433–2440, at p. 2435–2436.

[72] L.I. Tennen, "Enterprise Rights…", *op. cit.*, p. 287. See also L.I. Tennen, "Second Commentary on Emerging System of Property Rights in Outer Space", Proceedings of the United Nations/Republic of Korea Workshop on Space Law "United Nations treaties on outer space: actions at the national level", Daejeon, Republic of Korea, 3–6 November 2003, 2004, ST/SPACE/22, p. 343.

rights are not legally covered by domestic law. The IISL reached this position on the basis of the combined application of Articles II and VI OST:

> Article II of the 1967 Outer Space Treaty states that "Outer space, including the Moon and other celestial bodies, is not subject to national appropriation by claim of sovereignty, by means of use or occupation, or by any other means." The object and purpose of this provision was to exclude all territorial claims to outer space, including the Moon and other celestial bodies...
>
> Article VI of the Outer Space Treaty provides that "States bear international responsibility for national activities in outer space, including the Moon and other celestial bodies, whether such activities are carried on by governmental agencies or by non-governmental entities", *that is, private parties*, and "for assuring that national activities are carried out in conformity with the provisions set forth in the present Treaty". Article VI further provides that "the activities of non-governmental entities in outer space, including the Moon and other celestial bodies, shall require authorization and continuing supervision by the appropriate State Party to the Treaty."
>
> Therefore, according to international law, and *pursuant to Article VI, the activities of non-governmental entities (private parties) are national activities. The prohibition of national appropriation by Article II thus includes appropriation by non-governmental entities* (i.e. private entities whether individuals or corporations) since that would be a national activity. *The prohibition of national appropriation also precludes the application of any national legislation on a territorial basis to validate a 'private claim'*. Hence, it is not sufficient for sellers of lunar deeds to point to national law, or the silence of national authorities, to justify their ostensible claims. *The sellers of such deeds are unable to acquire legal title to their claims. Accordingly, the deeds they sell have no legal value or significance, and convey no recognized rights whatsoever.*
>
> *According to international law, States party to a treaty are under a duty to implement the terms of that treaty within their national legal systems.* Therefore, to comply with their obligations under Articles II and VI of the Outer Space Treaty, *States Parties are under a duty to ensure that, in their legal systems, transactions regarding claims to property rights to the Moon and other celestial bodies or parts thereof, have no legal significance or recognised legal effect*[73](Emphasis added).

In 2009, "in view of recent misleading views and discussions on this subject in the press", the IISL returned to the issue when, in a later statement, it not only reiterated its above position but went further, noting that existing space law does not specifically address the issue of exploitation of space resources and that a dedicated legal regime is needed for this purpose:

> ...any purported attempt to claim ownership of any part of outer space, including the Moon and other celestial bodies, *or authorization of such claims by national legislation, is forbidden as following from the explicit prohibition of appropriation, and consequently is prohibited and unlawful. Since there is no territorial jurisdiction in outer space or on celestial bodies, there can be no private ownership of parts thereof, as this would presuppose the existence of a territorial sovereign competent to confer such titles of ownership.*
>
> ...
>
> *At present, international space legislation does not include detailed provisions with regard to the exploitation of natural resources of outer space*, the Moon and other celestial

[73] IISL, Statement by the Board of Directors of the International Institute of Space Law (IISL) On Claims to Property Rights Regarding The Moon and Other Celestial Bodies, in: https://iisl.space/statement-by-the-iisl-board-of-directors-on-claims-to-property-rights-regarding-the-moon-and-other-celestial-bodies/ (last visit: 18.04.2023).

> bodies, although it does set down a general framework for the conduct of all space activities, including those of private persons and companies, with respect to such natural resources.
>
> *The IISL is of the opinion that a specific legal regime for the exploitation of such resources should be elaborated through the United Nations, on the basis of present international space law*, for the purposes of clarity and legal certainty in the near future. The IISL will continue to play an active role in any such discussions as they develop.[74]

Finally, it is worth noting that in the aforementioned monograph by R.J. Lee, the view is expressed that the relevant provisions of the 1979 Moon Agreement should be taken into account in attempting to interpret Article II OST, despite the fact that the said Agreement has so far been ratified by only a few States:

> Further, it may be useful to consider the relevant provisions of the Moon Agreement, because although it has not received widespread acceptance in the international community, its provisions may provide some guidance in the interpretation of Article II of the Outer Space Treaty, to which the Moon Agreement is intended to be an extension and thus complementary.[75]

It follows from the foregoing that the prevailing theoretical common ground is the broad interpretation of the concept of "national appropriation" under Article II of the OST—namely, that it also encompasses acts by private entities. This interpretation further complicates the application of existing legal norms to the exploitation of space resources. It is therefore not coincidental that current discussions on this issue within international *fora* attract growing interest from the global space community, as will be demonstrated below.

***

It is evident that the theoretical discussions surrounding the issue of space resources outlined herein ultimately lead to a deadlock. This is because such discussions, in practice, prove wholly ineffective in arriving at a persuasive and widely accepted interpretative model of current international space law—one that would provide the necessary legal framework to permit the exploitation of space resources. Furthermore, it has become clear that the central issue lies in the legality of the exploitation (rather than mere exploration) of space resources, insofar as such exploitation does not constitute simple use but results in their depletion. In any case, it appears that the combined application of Articles I and II of the 1967 Outer Space Treaty (freedom of use and exploration *versus* the principle of non-appropriation) creates, at least *prima facie*, an insurmountable legal barrier to the use and exploitation of space resources. Moreover, this combined application, in conjunction with Article III of the OST, leads to the conclusion that any viable legal solutions must emerge from international law—especially considering the nature of outer space as *res communis omnium*.

---

[74] IISL, Statement of the Board of Directors of the International Institute of Space Law (IISL), 22 March 2009, in: https://iisl.space/further-statement-by-the-iisl-board-of-directors-on-claims-to--lunar-property-rights/ (last visit: 18.4.2023).

[75] Lee, *op. cit.*, p. 167–168.

Chapter 4
Uncertainty of the Next Step #2: Reconciling State Perspectives Within the UNCOPUOS

In this part, the discussions, negotiations and different perspectives of States on the issue of mineral space resources in the framework of the United Nations Committee on the Peaceful Uses of Outer Space (UNCOPUOS, COPUOS) are approached. This Committee was established in 1959 by the United Nations General Assembly,[1] of which it is a subsidiary body. It consists of a Scientific and Technical Subcommittee and a Legal Subcommittee. All these bodies meet annually.

Over the years, COPUOS has been instrumental in the negotiation and adoption of the five international space-related treaties, while its role in promoting non-binding principles on critical issues of space activities has been crucial.[2] Its work has been vital in promoting international cooperation in the context of space activities, and many issues have emerged and evolved through the interaction of its Member States.

COPUOS is considered to be the most important forum for space issues. It is not the only forum that has dealt with the issue of space resources—The Hague International Space Resources Governance Working Group could also be mentioned in this context. However, the uniqueness of COPUOS lies in the fact that it is the global platform for the discussion of space issues—it is the only body that brings the "UN spirit" to space matters.[3]

---

[1] RES 1472 (XIV), International Co-operation in the Peaceful Uses of Outer Space, General Assembly, 14th session, 12.12.1959.

[2] Such as: Principles Governing the Use by States of Artificial Earth Satellites for International Direct Television Broadcasting, A/RES/37/92 of 10 December 1982; Principles relating to remote sensing of the Earth from outer space, A/RES/41/65 of 3 December 1986; Principles Relevant to the Use of Nuclear Power Sources in Outer Space, A/RES/47/68 of 14 December 1992; Space debris mitigation guidelines, UNCOPUOS, A/RES/62/217 of 22 December 2007; Guidelines for the Long-term Sustainability of Outer Space Activities, A/RES/74/82 of 13 December 2019.

[3] See, for instance, A/RES/79/87 of 4 December 2024, in the context of which COPUOS is referred to as "the unique platform at the global level for international cooperation in space activities".

G. D. Kyriakopoulos, *Space Resource Utilization and Exploitation: The Suspended Step of International Law*, Space and Society,
https://doi.org/10.1007/978-3-031-96336-0_4

In the following, the work of both COPUOS and, in particular, its Legal Subcommittee, on the issue of space resources will be presented in detail, from the introduction of this issue on the LSC agenda to the creation and launch of the Working Group on Legal Aspects of Space Resource Activities. Through this tour, the interactions among States Parties, their differing views and positions, and the challenges in establishing common ground on the issue under consideration will be clearly highlighted. The persistent impasses highlight the need for global consensus—especially within COPUOS, which, along with its subsidiary bodies, operates by consensus.

This section is devoted exclusively to COPUOS. The work of another important forum on this topic, The Hague International Space Resources Governance Working Group, will be examined mainly in terms of its outcome (the Building Blocks for the Development of an International Framework on Space Resource Activities) and will therefore be addressed in the next part.

4.1 55th Session of the LSC (4–15 April 2016): Introducing a New Agenda Item

During the 55th Session of the COPUOS Legal Subcommittee (LSC), the Belgian delegation, in its statement under agenda item "General Exchange of Views", referred to "the legislative and governmental initiatives taken by States in the area of exploitation of natural resources of outer space". Belgium focused on the fact that said initiatives put "fundamental questions regarding the future of space law and global economy to which the exploitation of space resources could contribute substantially". On this point, Belgium summed up its position as follows:[4]

1. Outer space natural resources cannot be appropriated through extension of national jurisdictions. Such appropriation, besides its incompatibility with international law, would mean risk of global economic imbalance because of exclusivity claimed by some to the detriment of others. Even if such an imbalance has not been envisaged in the short or medium term, any precedent which will be derived from any national claim over such resources could determine virtually irrevocably an advantageous position or a position of disadvantage of states regarding these resources and their derived products.
2. These resources, on the other hand, could be usefully and appropriately used for the benefit of the community of nations, taking into account the legitimate interests of investors as well as the interests of the least favored countries, including in particular developing countries.

[4] See the relevant digital recording of the Session: 2016-04-05T10:00:00 – 2016-04-05T13:00:00 (around 1:00), at https://www.unoosa.org/oosa/audio/v3/index-staging.jspx. See also in: Irm. Marboe, "The End of the Concept of Common Heritage of Mankind", Proceedings of the International Institute of Space Law, Vol. 59, 2016, p. 230 et seq.

3. The role of outer space powers, and their industry in particular, should be recognized as the key element of the international regime to be established. Accordingly, the model adopted by the Montego Bay Convention on the International Law of the Sea could provide valuable indicators as to the role of all stakeholders, whether State or private, when it comes to implementing this regime.
4. Such a regime does not preclude State intervention through national laws. Quite the contrary, States should have a fundamental role in the implementation.
5. While Belgium has a natural and logical preference for an international solution to be elaborated and adopted in the framework of the 1979 Moon Agreement, in application of Article 11 in particular, it did not preclude the idea of an international instrument open to all States establishing an *ad hoc* regime responding to the characteristics mentioned.

Belgium's declaration was a key position at the time, at a later stage than the adoption of the US Act, because it established a number of crucial parameters for the question of the exploitation of space resources. Belgium implicitly, but clearly, emphasized the application of international law, as opposed to national laws attempting to regulate the issue, thus pointing out the "risk of global economic imbalance because of exclusivity claimed by some to the detriment of others". Belgium also referred to an "international regime" that would not ignore investors or "space powers", but would ensure that the use of space resources would be for the "benefit of the community of nations". Last but not least, Belgium introduced the idea of an *ad hoc* international regime as an alternative to the resource exploitation regime provided for in the 1979 Moon Agreement.

4.1.1 Discussion Under Agenda Item "Status and Application of the Five United Nations Treaties on Outer Space"

Of particular interest was the exchange of views that took place in relation to the adoption by the United States of domestic legislation encouraging the private exploitation of the resources of celestial bodies [U.S. Commercial Space Launch Competitiveness Act (H.R. 2262, 25.11.2015)].

During the discussion on this agenda item, a significant number of States expressed their fear that national legislation adopted by certain States to unilaterally secure private rights to space resources that can be extracted from celestial bodies could lead to claims of sovereignty in outer space and to forms of national appropriation of those resources, in violation of fundamental principles of space law. The discussion focused mainly on the US Act, but also on the Luxembourg Space Resources Initiative.

In connection with the US Act, the representative of Russia questioned whether it was legal and ethical for a country to adopt domestic legislation on the use and exploitation of mineral resources from asteroids and other celestial bodies. He asked the LSC to define the framework for discussion on this issue. For Russia, the problem of interpreting the relevant provisions of the Treaties was of great importance and should be resolved at the universal level.

The United States responded that the existence of national laws licensing private activities in space and protecting property rights does not in itself constitute a violation of the OST, since no private party has been authorised to mine or exploit the resources of the Moon or other celestial bodies. They also considered that the LSC should discuss evaluation criteria in light of the new circumstances. Finally, the US assured that any future action would be taken in accordance with the international obligations of the United States.

Belgium (one of the countries that have ratified the Moon Agreement) noted that investors should at some point receive comprehensive answers but considered that Russia had been absolutely precise in its arguments, so the question remains what the future holds on this issue. During an informal meeting of the European Space Agency (ESA) Member States, the Belgian representative stressed that silence was not a solution to the actions of some States, such as the United States, to adopt national legislation on the commercial exploitation of space resources and considered that it was also in the interests of the United States to cooperate on this issue. According to Belgium, there should be a separate item for discussion on the LSC agenda on the commercial exploitation of space. Furthermore, Belgium stressed again that the exploitation of the resources of the celestial bodies should be for the benefit of all State and that, to that end, the provisions of the 1982 Convention on the Law of the Sea, as regards the international seabed, could serve as a model.

The delegation of Luxembourg, for its part, presented the website *www.spaceresources.lu*, in the context of creating a legal framework to facilitate private investors in space activities.

On the contrary, the representative of Austria (that has also ratified the Moon Agreement) stressed that States that adopt national rules should do so in compliance with international norms. It went on to commend the Moon Agreement, noting that said agreement facilitates international cooperation on the use of outer space and provides the necessary basis for discussion to work towards a multilateral solution.

Netherlands referred to the establishment of The Hague International Space Resources Governance Working Group, whose purpose was to search for a governance framework for space resources, with the participation of non-governmental stakeholders and academic institutions.

Mexico, another country that has ratified the Moon Agreement, focused on Article II of the OST, but also on the fact that Article 11 of the MUN essentially allows for the commercialisation of space resources. He emphasised that space resources are a common heritage of mankind, a concept that also exists in the 1982 Convention on the Law of the Sea, and generally noted that it supported the positions of Belgium.

Finally, the representative of the International Institute of Space Law (IISL, observer) reiterated the Institute's (balancing) position on the issue, according to which, given the absence of an explicit prohibition of appropriation of resources in the OST, it should be accepted that the utilization of space resources is allowed. This view, of course, is independent of claims of sovereign rights over celestial bodies.

4.1.2 Discussion Under Agenda Item: "Proposals to the Committee on the Peaceful Uses of Outer Space for New Items to be Considered by the Legal Subcommittee at its Fifty-Sixth Session"

In the context of this agenda item, following a Belgian initiative, which was supported by several other delegations, such as Austria, the Netherlands, Mexico, Iran, Greece and the Russian Federation, a proposal for the inclusion of a new agenda item on a "general exchange of views on potential legal and economic models for space [mineral] resources exploitation" was finally adopted by the Subcommittee. According to Belgium, a discussion on the general principles of a legal framework applicable to resource activities was necessary even before any such activity becomes feasible. In this respect, the proposed agenda item could provide the context for an exchange of views among States and other stakeholders that participate in the works of the LSC as observers.[5]

After discussion, the LSC indeed adopted the aforementioned proposal and decided that a new single issue/item for discussion, entitled "*General exchange of views on potential legal models for activities in exploration, exploitation and utilization of space resources*", should be included on the agenda of the LSC at its 56th Session. The LSC also agreed that "the inclusion of that item would provide an opportunity for a constructive, multilateral exchange of views on such activities, including their economic aspects, among States members and permanent observers of the Committee".[6]

Space resources were thus introduced into the LSC agenda, where they were to become a key topic of debate.

4.2 56th Session of the LSC (27 March–7 April 2017):[7] Initial Approaches to the New Agenda Item

4.2.1 Discussion Under Agenda Item: "General Exchange of Views"

In the context of this agenda item, representatives of several Member States referred, directly or indirectly, to the issues raised under the (new) Agenda Item 14 "General exchange of views on potential legal models for activities in

[5] See Marboe, op. cit., p. 234 et seq., as well as the relevant digital recordings of the 55th Session of the LSC, in: https://www.unoosa.org/oosa/audio/v3/index-staging.jspx.

[6] Report of the Legal Subcommittee on its fifty-fifth session, held in Vienna from 4 to 15 April 2016, UN Doc. A/AC.105/1113, 27 April 2016, para 250.

[7] See the relevant digital recordings of the 56th Session of the LSC, in: https://www.unoosa.org/oosa/audio/v3/index-staging.jspx. Also see Report of the Legal Subcommittee on its fifty-sixth session, held in Vienna from 27 March to 7 April 2017, A/AC.105/1122.

exploration, exploitation and utilization of space resources":[8] Costa Rica (on behalf of the Group of 77 + China) expressed the view that there should be a universal and equitable access of States to space and a rational use of space resources. This view was shared by Ecuador, Indonesia and Pakistan. Venezuela (on behalf of the GRULAC) noted that current technological advances were unthinkable at the time of the conclusion of the "space" treaties, hence the vital need to update space law in order to achieve the necessary balance. China, also advocating the need for a reasonable exploitation regime, has further stressed the need to adopt new international rules to regulate emerging private activities in space. For its part, the Russian Federation stressed that if unilateral approaches to exploiting space resources were to circumvent international rules, nothing less than chaos could result. It further added that the basis for a pragmatic approach to such activities remains the 1967 Outer Space Treaty. Canada, on the other hand, focused mainly on the extraordinary advantages that space innovation brings, arguing that it would be particularly important to have a common legal foundation based on all four 'space treaties' (thus excluding the Moon Agreement). Finally, the United States stressed that the rise of private actors was not a new phenomenon in space (thus implicitly arguing that the demand for new arrangements was not pressing) but it was willing to participate in the discussion under agenda item 14.

4.2.2 Discussion Under Agenda Item: "General Exchange of Views on Potential Legal Models for Activities in Exploration, Exploitation and Utilization of Space Resources"

This was the first time that this item was on the LSC Agenda, but it attracted the general interest of the Member States. A substantive discussion was carried out, but without strong controversy, and all Member States that took the floor welcomed the item on the LSC agenda and were in favor of keeping it for the next session. In particular:

Costa Rica (Group 77 + China) argued in favor of the need for a new legal regime that would provide the required security, and further stressed that developing countries should not be excluded from the exploitation of the outer space. China referred, *inter alia*, to the balance to be achieved between national interests and the interests of the international community and pointed out that the broadest possible consensus should be sought. For its part, the United States clearly favored unilateral national exploitation activities in outer space: its representative recalled that, as

[8] A/AC.105/C.2/L.299, 27 January 2017.

early as 1979, the official US position was that non-appropriation concerns resources *in situ*, so resources that are extracted and removed are not covered by this principle. They further expressed their position that their 2015 national legislation did not violate their international obligations and was in harmony with international space law.

The Russian Federation noted that the relevant discussion was of a theoretical nature and stressed that its basis could only be in the direction of international regulations. Finally, it observed that the Outer Space Treaty did not refer *expressis verbis* to the commercial exploitation of celestial bodies, so an interpretation of the relevant terms was particularly crucial. However (according to the RF), this interpretation could not be unilateral.

Germany has followed a similar, but more eloquent, approach: it has advocated a multilateral international framework for regulating new activities and, interestingly, has explicitly focused on the need for an interpretative clarification of the relationship between Articles I (freedom of exploration and use of Outer Space) and II (non-appropriation principle) of the OST. It is significant that Germany had clearly taken the view that the principle of non-appropriation is not covered by the freedom of exploration and use, with the obvious consequence that it operates as a limitation on this freedom. Along essentially the same lines, Italy also opposed the creation of national, unilateral legislation on the exploitation of outer space, stressing that such activities should be carried out in accordance with the basic obligations laid down in the space treaties.

Venezuela's (GRULAC) position was in the same line, underlining that the freedom of use of outer space was limited by the principle of non-appropriation—and that COPUOS should contribute to avoiding interpretations of a unilateral nature. France also stressed that priority should be given to an international multilateral approach.

In an attempt to create a draft common ground that could serve as a starting point for discussion, Belgium submitted a paper entitled "Contribution from Belgium to the discussion under UNCOPUOS Legal Subcommittee on item 'General exchange of views on potential legal models for activities in exploration, exploitation and utilization of space resources'" (A/AC.105/C.2/2017/CRP.19). In this document, Belgium stated, *inter alia*, that the prospect that some States with the highest possible technological development would benefit exclusively from space resources did not seem to be in line with the usual interpretation of the terms of the 1967 Outer Space Treaty and the principle of good faith. Belgium also stressed that, despite its view that the 1979 Moon Agreement would be a good starting point for the establishment of a regime for space resources, it was not opposed to the idea of a new treaty text, which should, however, be based on a number of principles compatible with the Outer Space Treaty, such as equitable access to the resources of outer space and the granting of rights and interests to developing countries and 'pioneering' States alike.

The States Parties agreed to keep this item on the agenda of the LSC for the next LSC Meeting (2018).

4.3 57th Session of the LSC (9–20 April 2018):[9] A Greek Proposal to Discuss the Topic in the Context of the Working Group on the Status and Application of the Five United Nations Treaties on Outer Space

4.3.1 Discussion Under Agenda Item "General Exchange of Views on Potential Legal Models for Activities in Exploration, Exploitation and Utilization of Space Resources"

During this year's debate on the issue, Germany made a statement clearly in favor of a collective exploitation of space resources, referring to Article 11(5) of the Moon Agreement and the need for an international regulation. Austria also stressed the need for a multilateral approach. Belgium presented a written memorandum, while Mexico proposed to hold a conference to review the space treaties.

At that session, the Greek delegation made a statement in which it discussed the nature of existing space law as a factor to be considered in the context of the debate on the initiation of activities for the exploitation of space resources. Greek also proposed the establishment of a working group to facilitate consideration of this matter.

4.3.2 Discussion Under Agenda Item "Proposals to the Committee on the Peaceful Uses of Outer Space for New Items to be Considered by the Legal Subcommittee"

When, during the second week of the session, proposals for new items to be considered by the Sub-Committee were discussed under agenda item 16, the Greek delegation finally proposed that agenda item 15 be dealt with by the existing working group on the status and application of the five United Nations treaties. The Netherlands and the United States were skeptical about this, however Brazil supported the Greek proposal and suggested that the countries that agreed should work with Greece to present a proposal at the 58th Session of the LSC (2019) on the feasibility of including the issue in the agenda of that Working Group. There was a broad support for this proposal, with Belgium, Mexico, the Czech Republic, Australia, Luxembourg, Mexico, Australia and China in favor.

[9] See the relevant digital recordings of the 57th Session of the LSC, in: https://www.unoosa.org/oosa/audio/v3/index-staging.jspx, as well as Report of the Legal Subcommittee on its fifty-seventh session, held in Vienna from 9 to 20 April 2018, A/AC.105/1177.

4.4 58th Session of the LSC (1–12 April 2019):[10] A Greek-Belgian Proposal for the Establishment of an *Ad Hoc* Working Group

4.4.1 Discussion Under Agenda Item "General Exchange of Views on Potential Legal Models for Activities in Exploration, Exploitation and Utilization of Space Resources"

Following the work of the previous session, a draft document was prepared at the initiative of Greece, which, after informal consultations and exchanges of views, received comments and additions from Belgium, while further comments were incorporated from Germany, which expressed its interest in supporting this initiative. The draft finally took the form of a joint Greek-Belgian proposal, entitled "Proposal for the establishment of a working group for the development of an international regime for the use and exploitation of space resources—Working paper by Belgium and Greece" (A/AC.105/C.2/L.311/4.3.2019, hereinafter referred to as "L 311").

Document L 311 highlighted the principles of international space law that should govern the future use of the resources of celestial bodies, particularly: (1) that the exploration and use of outer space is the province of all humankind; (2) that such exploration and use is governed by international law; and (3) that international space treaties require relatively broad international cooperation. Based on these principles, the document proposed the establishment of an *ad hoc* working group within the LSC with the mandate to examine and report on the issues and questions raised in connection with the future use and exploitation of space resources, as detailed in the document. It should be noted that the co-signatories of the proposal (Greece and Belgium) explicitly considered COPUOS as the main forum for discussions and negotiations between States on the future regime for the use of space resources, which should be international in character and should focus on the main objectives set out in Article 11, para. 7 of the 1979 Moon Agreement, irrespective of the status of ratification.

The 58th Session of the LSC was largely focused on debating L 311. The proposal was presented by the Greek and Belgian delegations under both item 14 ("General exchange of views on potential legal models for activities in exploration, exploitation and utilization of space resources") and item 15 ("Proposals to the Committee on the Peaceful Uses of Outer Space for new items to be considered by the Legal Subcommittee at its fifty-ninth session") on the agenda of the LSC, but also at informal meetings of the Member States of the European Space Agency (ESA) and the European Union.

---

[10] Digital recordings of the 58th Session of the LSC, in: https://www.unoosa.org/oosa/audio/v3/index-staging.jspx; Report of the Legal Subcommittee on its fifty-eighth session, held in Vienna from 1 to 12 April 2019, A/AC.105/1203.

In particular:

During the discussion under agenda item 14 ("General exchange of views on possible legal models for activities in the exploration, exploitation and use of space resources"), the proposal received broad support from Austria, China, Czech Republic, France, Germany, Brazil, Russia and Mexico. However, reservations were expressed, mainly by the Netherlands (it was in favor of an approach that would not identify an exclusive forum for discussion and of limited duration), the United States (arguing that the exploitation of space resources was still at an early stage and, therefore, there was no reason to establish a working group on this issue) and Luxembourg (not opposing in principle to the creation of the working group, arguing, however, that there were important policy issues beyond the legal dimension of the matter).

In the absence of the necessary consensus, and having taken note of the comments made by States, the Greek and Belgian delegations prepared an *Addendum* on the Working Methods and Work Plan of the Working Group proposed under document A/AC.105/C.2/L.311 entitled "Proposal for the establishment of a working group for the development of an international regime for the utilization and exploitation of space resources", which was codified as document A/AC.105 /C.2/2019/ CRP.22/8.4.2019 (hereinafter referred to as CRP 22) and introduced for discussion under agenda item No. 15: "Proposals to the Committee on the Peaceful Uses of Outer Space for new items to be considered by the Legal Subcommittee at its fifty-ninth session". CRP 22, in response to the above comments by States, referred to a method and programme of work for the working group for the period 2020–2022.

On the basis of CRP 22, Egypt, Iran and the United Arab Emirates were added to the States supporting the proposal, while the Russian Federation, which continued to support the proposal, introduced a new element to the debate, namely the involvement of the Scientific and Technical Subcommittee (STSC) of the COPUOS (a proposal to which Mexico and Iran agreed).

The United States, on the other hand, remained skeptical, mentioning that the proposal took for granted the need for a legal regime to address the issue of exploration, utilization and exploitation of space resources. Along the same lines, Canada considered that the proposal needed further elaboration, whereas the United Kingdom argued that this activity was "immature", given the lack of sufficient development of the necessary technology, and called for further consultations. The Netherlands remained on the same line and asked for further discussion, as also did Luxembourg.

At this point, China asked the Secretariat whether it would be feasible to have an informal mechanism that would allow States to discuss in depth the modalities for the establishment of a working group.

In response, the Secretariat emphasised that (1) In line with the practice followed for the establishment of the Working Groups on "Space and Global Health" and the "Space 2030 Agenda" (under the Scientific and Technical Sub-Committee, STSC), the Working Group could be established (as a "shell") and then its Bureau would coordinate consultations to define its precise mandate and work plan, (2) There was an earlier occasion (1996) when the (then) President of the LSC asked the (then)

Secretary to act as a moderator to conduct consultations to define new items for the LSC agenda. This was a "formal informal" process.

On the basis of the new reactions and in order to facilitate consensus, the two delegations (Greece-Belgium) proceeded to draft a new (2nd) revised document entitled: "Proposal for Working Methods and Work Plan of the Working Group on Legal Aspects of the Exploration, Utilisation and Exploitation of Space Resources (with reference to document A/AC.105/C.2/L.311)", which was codified as document: A/AC.105/C.2/2019/CRP.26/11.4.2019 (hereinafter referred to as CRP 26). According to that document, the proposed working group to be established now referred to "legal aspects of the exploration, use and exploitation of space resources" instead of "regime", to show the flexibility needed to reach consensus.

In the same vein, the detailed work plan of CRP 26 referred to "potential legal gaps" and "possible development of international legal framework and international cooperation mechanisms, through binding or non-binding instruments".

Following the submission of CRP 26, South Africa was added to the supporters of the proposal. Luxembourg, however, insisted that it was in favor of a holistic approach to the issue, not limited to its legal aspects, but taking into account its political and technological dimensions. The Netherlands, for its part, reiterated that the mandate of the proposed working group needed further discussion.

During the discussion, Russia, China, Mexico, South Africa and Iran supported the *interim* solution (as proposed by the Secretariat) of first establishing the Working Group and then determining its mandate and work plan. The United States, however, did not agree with this alternative and favored further informal consultations along the lines previously mentioned by the Secretariat.

At the suggestion of the Chair, the Secretariat re-explained the rationale for "formal informals" in line with the practice established in 1996–1997, an idea supported by the US, Canada, the UK, the Netherlands and Japan, while China, Russia and Mexico insisted that the establishment of the Working Group for 1 year and the subsequent definition of its mandate and programme was the optimal solution.

4.4.2 *The "Scheduled Informal Consultations"*

Faced with an impasse, the Chair put forward a compromise proposal, which was unanimously accepted by States. According to said proposal, scheduled informal consultations were to be held at the 59th Session of the Subcommittee, in 2020. Those scheduled informal consultations were to be conducted with interpretation services and be coordinated by a moderator assisted by the Secretariat. The moderator was to be nominated by Belgium and Greece (at the 62nd Session of the COPUOS, in 2019).[11]

---

[11] Report of the Legal Subcommittee on its fifty-eighth session, held in Vienna from 1 to 12 April 2019, 18 April 2019, A/AC.105/1203, para. 278.

The Subcommittee further agreed that "the aim of those consultations was to have a broad and inclusive exchange of views on the future deliberations concerning the exploration, exploitation and utilization of space resources, including the possible establishment of a working group under the relevant agenda item, taking into account possible future coordination with the Scientific and Technical Subcommittee, as appropriate".[12] In this context, "Belgium and Greece would submit a revised proposal for the establishment of a working group" under the relevant agenda item "for further consideration at the fifty-ninth session of the Subcommittee. The Subcommittee encouraged those two delegations to conduct consultations with interested delegations on the margins of the sixty-second session of the Committee".[13]

In fact, Belgium and Greece, at the 62nd session of the COPUOS (12–21 June 2019), have jointly nominated (1) Mr. Andrzej Misztal, Polish diplomat and former President of the LSC, as moderator; and (2) Mr. Steven Freeland, member of the Australian delegation to the NYE, Professor of International Law, as deputy moderator, in the framework of the aforementioned decision of the LSC. The Committee unanimously approved the appointment of the moderators. The moderators also undertook to present, during the intersessional period, a draft programme for the conduct of the "scheduled informal consultations", including substantive proposals for discussion and the rationale for these proposals. Member States will be invited to comment on these proposals.

In parallel with the appointment of the moderators, Greece and Belgium have continued their efforts to reach a consensus on the establishment of the working group and have held informal consultations with interested delegations in the margins of the session.

Finally, it should be noted that the United Arab Emirates submitted to the plenary its own proposal for the establishment of a working group on the use of space resources (A/AC.105/2019/CRP.17) "in support of the possible establishment of a working group related to the Space Resource Utilization". As stated in the text of the proposal, this document was a first contribution by the United Arab Emirates in the context of the informal consultations on this issue.

***

It follows from the foregoing that the joint Greek-Belgian proposal for the creation of a working group at the 2019 LSC session, although it did not achieve consensus, has nevertheless given a strong impetus to the international debate on the issue of space resources. As will be shown below, this has been ultimately confirmed by subsequent developments.[14]

---

[12] *Idem*, para. 279.

[13] *Idem*, para. 280.

[14] See in this respect, F. Xu & J. Su, "Towards a legal regime of benefits sharing for space mining: With some experience from the Area", Resources Policy, Volume 76, 2022, article id. 102627, https://doi.org/10.1016/j.resourpol.2022.102627, p. 4.

4.5 60th Session of the LSC (31 May–11 June 2021): The Establishment of the Working Group on Space Resources

4.5.1 Discussion Under Agenda Item 14 "General Exchange of Views on Potential Legal Models for Activities in Exploration, Exploitation and Utilization of Space Resources"

Since, in 2020, there was no Session of the LSC, in fact, due to the COVID pandemic, the above "scheduled informal consultations" could not take place. In September 2019, the moderators circulated the draft programme, on which they received responses from 14 Member States of the Committee. At the same time, they held informal discussions and consultations with various delegations during the intersessional period. Due to the COVID-19 pandemic, the 59th Session of the LSC was cancelled and, in accordance with the decisions and actions of COPUOS taken by written procedure, the "scheduled informal consultations" were finally held during the LSC meetings in 2021.

During the 60th Session of the LSC, the moderators held eight (8) rounds of "scheduled informal consultations". Three proposals were put before the LSC Member States:

1. A German-Finnish Proposal, which was an initiative of these countries and was eventually submitted as a joint proposal by a significant number of (European) States (Austria, Belgium, Czech Republic, Finland, Germany, Greece, Poland, Portugal, Romania, Slovakia and Spain), entitled: "The Establishment of a Working Group on Potential Legal Models for Activities in Exploration, Exploitation and Utilization of Space Resources" (A/AC.105/C.2/2021/CRP.22) (hereinafter "CRP 22"). The proposal stressed that the working group on space resources should be established "without further delay", while its purpose would be "to ensure that space resources activities are conducted in a safe, sustainable and peaceful manner, for the benefit and in the interests of all countries, irrespective of their degree of economic or scientific development, and in accordance with international law".[15]

It should be noted that CRP 22 explicitly acknowledged the importance of the Greek-Belgian proposal that had been presented and submitted at the 2019 Session, of which it was in a sense a natural continuation, since the introduction of the text states that: "This proposal builds upon previous working papers submitted on the issue of space resources that have been instrumental in advancing

[15] The Establishment of a Working Group on Potential Legal Models for Activities in Exploration, Exploitation and Utilization of Space Resources, Proposal submitted by Austria, Belgium, Czech Republic, Finland, Germany, Greece, Poland, Portugal, Romania, Slovakia and Spain, 8.6.2021, A/AC.105/C.2/2021/CRP.22.

the debate on the establishment of a Working Group on Space Resources, notably the joint proposal by Belgium and Greece...".
According to CRP 22, the purpose of the working group would be, on the basis of a 5-year work plan, to submit to the Legal Subcommittee:

> 1. principles and practical measures for space resources activities on celestial bodies, for possible adoption by the United Nations General Assembly in a dedicated resolution; and
> 2. a report on the legal aspects of space resources activities, including an assessment of the expected benefits and means of the further development of an international framework for space resources activities as well as recommendations for further work at the Legal Subcommittee on this topic.

In line with the above, this proposal considered that, within the scope of the working group, this would include, *inter alia*, an assessment of the establishment of an "international framework" for space resources activities.

2. A second draft was submitted by the Russian Federation (The Establishment of a Working Group on Potential Legal Models for Activities in Exploration, Exploitation and Utilization of Space Resources, A/AC.105/C.2/2021/CRP.26). That draft placed particular emphasis on the legal dimension and regulation of the issue of space resources, listing as the subject of the working group being established:

> a. Studying applicable norms of international law defined, primarily, by the five United Nations treaties on outer space.
> b. Identifying possible gaps in applicable international legal norms, guidelines and rules regarding regulation of activities related to the exploration, exploitation and utilization of space resources.
> c. Developing principles (guidelines) and rules (proposals/recommendations) for the regulation of activities on exploration, exploitation and utilization of space resources, which could become elements of a new international treaty in the future.[16]

3. Finally, a third draft was submitted by China (The Establishment of a Working Group on Potential Legal Models for Activities in Exploration, Exploitation and Utilization of Space Resources, A/AC.105/C.2/2021/CRP.18). The Chinese draft clearly supported the establishment of the working group, which should "be guided in its work by international law, primarily international space law with OST as an example, and make efforts to safeguard its central role in the global governance of outer space activities. Any envisaged outcome of the Working Group, in whatever form, shall be consistent with the fundamental principles of international space law". The ultimate purpose of the working group, according

[16] The Establishment of a Working Group on Potential Legal Models for Activities in Exploration, Exploitation and Utilization of Space Resources, Proposal submitted by the Russian Federation, 8.6.2021, A/AC.105/C.2/2021/CRP.26.

to this draft, was "to study the legal framework for the further regulation of space resource activities, and design the structure for the final report of the Working Group".[17]

Countries such as the United States or the Netherlands, which had previously (2019) expressed strong skepticism about the creation of the working group, adopted a more positive attitude at this session (through statements during the discussion of agenda item 14 and during the scheduled informal consultations). China and Brazil stressed that the first priority should be to reach an agreement on the creation of the working group. Moderator Professor Steven Freeland (Australia) noted a "growing consensus" on this issue. Canada, a country among the skeptics, focused on the need to determine what the expected outcome was and the need for a solid technical input into the discussions. New Zealand was also positioned in the same direction.

4.5.2 The Co-Moderators' Proposal

I. Based on the processes and discussions, in the context of the "scheduled informal consultations", the moderators presented their proposal (Co-Moderators' Proposal, 4 June 2021). The main points discussed were as follows:

   (A) *As to the name of the working group*, between the "broad" (Working Group on Potential Legal Models for Activities in Exploration, Exploitation and Utilization of Space Resources) and the "short" version (Working Group on Space Resources). Brazil proposed an "intermediate" version: Working Group on Space Resources and Related Activities. China and Russia were in favor of the 'broad' name, for clarity.

   (B) *As to the duration of the relevant mandate:* the proposal referred to an "initial period of 5 years". Canada and the United Kingdom requested that the period should not be considered "initial" and therefore not renewable.

   (C) *As regards the existing legal framework:* Japan proposed that instead of referring to the five United Nations treaties on outer space, the reference should be to "treaties and other relevant instruments".

   (D) *On the benefits of developing an ad hoc legally binding international framework:* The United States and Australia expressed the view that reference to a "legally binding" international framework should be avoided. Israel agreed with this view.

   (E) *Regarding the participation of the Scientific and Technical Subcommittee:* Canada and Japan considered it desirable to involve the COPUOS Scientific and Technical Subcommittee in the whole process, to (also) discuss the technical aspects of the issue.

[17] The Establishment of a Working Group on Potential Legal Models for Activities in Exploration, Exploitation and Utilization of Space Resources, Proposal submitted by China, 1.6.2021, A/AC.105/C.2/2021/CRP.18.

(F) According to the proposal, the first task of the working group would be to collect information and data: 'Collect relevant information concerning activities in exploration, exploitation and utilization of space resources'. The United States requested the addition: 'including potential models for benefit sharing [of space resources]'.[18]

II. In view of the scheduled informal consultations of 7.6.2021, the moderators presented a revised proposal (Revised Co-Moderators' Proposal, 7 June 2021). In this proposal, the [above, under (F)] collection of information was now more specifically defined ("Collect relevant technical, scientific and legal information concerning activities in the exploration, exploitation, and utilization of space resources on an ongoing basis taking into account their innovative and evolving nature"), the reference to the existing legal framework was now defined as "in particular the Outer Space Treaty and the other relevant United Nations treaties and instruments adopted within the framework of the Committee", instead of "in particular the five United Nations treaties on outer space and other relevant instruments..." (Proposal of 4 June 2021). Further, reference was made to the need to create a "framework", without further specification (legal, binding), noting only, in an attempt to reconcile opposing views, that the mandate of the working group to be established would be, *inter alia*, to "identify areas where further development of the framework might be helpful and assess the benefits of doing so, including by way of potential legally binding and/or other instruments".[19]
Finland supported the revised proposal. Australia insisted that the terms "applicable"—"legal" be removed as identifiers of the existing legal framework [in the text: "Study the applicable existing legal framework for such activities ("framework")..."]. Belgium remarked in this respect that the work was taking place within the COPUOS Legal Subcommittee. The Russian Federation proposed, by way of compromise, to use the term "regulatory" instead of "legal". The United States proposed the term "governance framework", with the agreement of Canada. Greece requested that the difference between "regulatory" and "governance" be defined. China, Mexico and Venezuela requested that the reference to "legal" remain. Belgium proposed the term "normative" as an alternative, which was agreed by the United States, but also by Germany.

III. On the basis of the above consultations, the moderators presented their second revised Co-Moderators' Proposal (Second Revised Co-Moderators' Proposal, 9 June 2021). It referred, *inter alia,* to an "existing normative framework", which included "in particular the Outer Space Treaty and other applicable United Nations treaties, also taking into account other relevant instruments, as appropriate".
Belgium supported the proposal, while China insisted that the "framework" should be defined as "legal". The United States stated that it could accept

[18] Co-Moderators' Proposal (4 June 2021), https://www.unoosa.org/oosa/en/ourwork/copuos/lsc/space-resources/scheduled-informal-consultations.html.

[19] Revised Co-Moderators' Proposal (7 June 2021), https://www.unoosa.org/oosa/en/ourwork/copuos/lsc/space-resources/scheduled-informal-consultations.html.

"legal" in relation to the existing framework, but not in relation to paragraph 3 of the proposal ("Identify areas where further development of the framework might be helpful..."). Germany referred to "existing legal framework", to which China agreed.[20]

IV. Finally, the Co-Moderators submitted their Third Revised Co-Moderators' Proposal (9 June 2021; final consultations), in which, by way of compromise, paragraph 2 of the "Mandate" referred to "existing legal and normative framework".

On the third paragraph of the section: "Terms of reference" ["The Working Group will be led by a Bureau and shall be supported by the Secretariat"], Greece proposed, instead of "Bureau", to refer to "Chair–Vice-chair". Canada and Mexico agreed to this proposal. Canada proposed that the two moderators, Andrej Misztal and Steven Freeland, should occupy the positions of Chair and Vice-Chair, respectively, in the working group to be established. This proposal was agreed to by several countries (Brazil, United Kingdom, Italy, Russia, Ukraine, Belgium) without any objection being raised.[21]

4.5.3 The Establishment of the Working Group

Finally, on 9 June, the Legal Subcommittee decided, on the basis of the moderators' reports on the progress of the scheduled informal consultations, to establish, on the basis of a 5-year work plan, the working group "under the agenda item on the general exchange of views on potential legal models for the exploration, exploitation and utilization of space resources", chaired by Mr. Andrej Misztal (Poland) and vice-chaired by Mr. Steven Freeland (Australia).[22]

The LSC welcomed the moderators' consolidated proposal for the establishment of a working group, which brought together many views expressed by delegations during the discussions on the mandate, responsibilities and working method of the working group. Further, the LSC requested the newly elected Chair and Vice-Chair of the Working Group to continue consultations, in the intersessional period, on the mandate, terms of reference and working method of the Working Group and to consult with the COPUOS Chair and the Secretariat on the scheduling of the 64th Session of the Committee to enable the Working Group to meet during that session. In this regard, the LSC recommended that the Commission should further consider the matter at its 64th Session.[23]

---

[20] Second Revised Co-Moderators' Proposal (9 June 2021), https://www.unoosa.org/oosa/en/ourwork/copuos/lsc/space-resources/scheduled-informal-consultations.html.

[21] Third Revised Co-Moderators' Proposal (9 June 2021- final consultations), https://www.unoosa.org/oosa/en/ourwork/copuos/lsc/space-resources/scheduled-informal-consultations.html.

[22] Report of the Legal Subcommittee on its sixtieth session, held in Vienna from 31 May to 11 June 2021, 2.8.2021, para. 255.

[23] *Idem*, paras.256–257.

The last revised proposal of the moderators (Third revised) is as follows:

Third Revised Co-Moderators' Proposal (9 June 2021-Final Consultations)

The Legal Subcommittee ("LSC") of the Committee on the Peaceful Uses of Outer Space ("Committee") establishes, for a period of 5 years, a Working Group under the agenda item "General exchange of views on potential legal models for activities in exploration, exploitation and utilization of space resources".

Mandate

The Working Group shall:

- Collect relevant information concerning activities in the exploration, exploitation, and utilization of space resources, including with respect to scientific and technological developments and best practices, taking into account their innovative and evolving nature, as a first step;
- Study the existing legal and normative framework for such activities, in particular the Outer Space Treaty and other applicable United Nations treaties, also taking into account other relevant instruments, as appropriate;
- Identify areas where further development of the framework might be helpful and assess the benefits of doing so, including by way of additional governance instruments in any appropriate form;
- Recommend an initial set of principles and practical measures for such activities to ensure that they are carried out in a safe, sustainable and peaceful manner and in accordance with international law for adoption by the Committee and possibly then by the United Nations General Assembly as a dedicated resolution;
- Identify areas for further work of the Committee and recommend next steps, which may include the development of potential models, rules and norms, for activities in the exploration, exploitation and utilization of space resources, including with respect to related activities and benefit sharing.

Terms of Reference

- The Working Group reports to the LSC;
- The meetings of the Working Group are guided by the rules of procedure, methods of work and established practices of the Committee, including with respect to contributions by permanent observers;

- The Working Group will be led by a Bureau and shall be supported by the Secretariat;
- The Working Group shall be open to all States members of the Committee, and participation of developing and emerging countries shall be encouraged.
- The Working Group shall take into consideration submissions on the issue of activities in the exploration, exploitation and utilization of space resources from States members and permanent observers of the Committee, as well as inviting non-governmental stakeholders, including from academia, civil society, technical experts, institutional and private actors to share their views in accordance with the rules of procedure, methods of work and established practices of the Committee;
- The Working Group and its members may avail themselves of previous work on the issue of activities in the exploration, exploitation and utilization of space resources, including but not limited to study reports, academic research and papers.

Work Plan and Methods of Work

- On the basis of the agreed mandate and terms of reference, the Working Group shall, in 2022, agree on its detailed work plan and methods of work. This shall include appropriate means to include the expertise and views of academia, civil society, technical experts, institutional and private actors as well as means of coordination with the Scientific and Technical Subcommittee;
- The foregoing shall not prevent the Working Group from undertaking initial administrative, information collection and stocktaking tasks as provided for in its mandate;
- The Working Group shall in its work avail itself of electronic means of communication whenever practicable and/or necessary.

In conclusion, the final establishment of the working group under the agenda item "General exchange of views on possible legal models for activities in exploration, exploitation and use of space resources" strongly promotes multilateralism and international cooperation in the context of the discussion of legal issues arising in relation to space activities. The fact that there was effectively no LSC session in 2020 due to the pandemic has obviously given the necessary time to allow the matter to come to maturity among the States members of the Committee. Despite the fact that several COPUOS Member States remained skeptical about the prospect of a (binding) legal regulation of activities related to space resources, the 60th Session of the LSC demonstrated that the need to explore a regime to regulate these activities, specifying the general legal framework provided by the UN space treaties, in particular the 1967 Outer Space Treaty, was particularly felt.

4.6 61st Session of the LSC (28 March–8 April 2022): A Greek Questionnaire

4.6.1 Discussion Under Agenda Item "General Exchange of Views on Potential Legal Models for Activities in the Exploration, Exploitation and Utilization of Space Resources"

During the session, Greece submitted, as its first contribution to the work of the Working Group, a questionnaire, which addressed "important aspects of the relevant debate and of the views expressed on the subject at the international level in recent years, considering that this may facilitate the exchange of views within the working group, but also assist the Chair and Vice-Chair in recording the positions of the members of the Subcommittee, with a view to exploring the necessary convergences and making the work of the Working Group meaningful and fruitful".[24] Said questionnaire has taken the form of a dedicated document (CRP 13)[25] and was also submitted in the context of Agenda item 15. The Greek questionnaire addressed the essence of the problem. Its purpose was to record the positions of the Committee's member States on the most important aspects of the debate on the exploration, utilisation and exploitation of space resources. Of course, States were always free to answer only in part or not at all. The questionnaire was as follows:

Questionnaire related to the discussion on potential legal models for the exploration, exploitation and utilization of space resources

1. Is the existing international legal framework for outer space activities (mainly the five United Nations Treaties) sufficient to regulate the exploration, exploitation and utilization of space resources (a) or is an *ad hoc* framework required (b)?

   1.1 If the answer to (a) is in the affirmative, which existing principles and provisions provide for an adequate regulation of the matter?

   1.2 If the answer to (b) is in the affirmative, what should be the main characteristics of *an ad hoc* framework?

[24] 61st Session of the Legal Subcommittee of COPUOS, Agenda item 15: General exchange of views on potential legal models for activities in exploration, exploitation and utilization of space resources, Statement by Greece: https://www.unoosa.org/oosa/audio/v3/index-staging.jspx.

[25] Working Paper submitted by Greece - Questionnaire related to the discussion of item 15 on potential legal models for the exploration, exploitation, and utilization of space resources, A/AC.105/C.2/2022/CRP.13, 28 March 2022, https://www.unoosa.org/oosa/oosadoc/data/documents/2022/aac.105c.22022crp/aac.105c.22022crp.13_0.html.

1.3 Further, this framework:

(a) Should it have the characteristics of a legal regime? If so, should the regime in question be regulated by international law or at the level of domestic law?
(b) Should it have the characteristics of an ex aequo et bono regime within the meaning of Article 38 para. 2 of the Statute of the International Court of Justice? If this should be the case, on which principles should such a regime be based?

2. Can the issue of exploration, exploitation and utilization of space resources be considered to fall within the scope of Article II of the Outer Space Treaty (non-appropriation principle)? In such a case, and bearing in mind that a framework for the regulation of resources is being explored.

2.1 Should Article II of the Outer Space Treaty be subject to a commonly accepted interpretation?
2.2 Should Article II of the Outer Space Treaty be amended?
2.3 Should there be a review of the concept of celestial bodies, so as to narrow the scope of Article II (e.g. as regards comets and asteroids)?

3. Should the issue of exploration, exploitation and utilization of space resources be considered with a view to an all-inclusive regulation, or should there be different approaches on a case-by-case basis (for example, in relation to orbits around the earth, to *in situ* resource utilization or to space mining for commercial exploitation)?
4. Should a legal framework for the exploration, exploitation and utilization of space resources promote an equitable access to said resources, for all States?
5. Could the Moon Agreement provide a suitable basis for the development of a legal regime for the exploration, exploitation and utilization of space resources?
6. In the context of the debate on a framework for the exploration, exploitation and utilization of space resources, should the interaction of this issue with the long-term sustainability of outer space activities be considered? In general, should the issue be considered in the light of environmental considerations, in accordance with the letter and spirit of Article IX of the Outer Space Treaty?
7. Could the existing international legal regime for mineral exploitation in the Area (international seabed) possibly serve as a model for the exploration, exploitation and utilization of space resources?
8. Are the technical and scientific aspects of the exploration, exploitation and utilization of space resources considered as factors that affect the development of a framework? If the answer is in the affirmative, what is the best way to make the relevant feedback available to the working group on space resources?

Some States welcomed the idea of a questionnaire (Finland, France). The Greek Delegation stated that the proposed questionnaire could serve as a working basis for the Chairs.[26]

Further, Luxembourg and the Netherlands, by common working paper, submitted to the LSC the outcome of the work of the Hague International Space Resources Governance Working Group (Building blocks for the development of an international framework on space resource activities).[27] Also, a group of (European) States (Austria, Belgium, Czech Republic, Finland, Germany, Luxembourg, Norway, Portugal and Romania) introduced a conference room paper "on the endorsement of the Work Plan for the working group established under the Legal Subcommittee agenda entitled "General exchange of views on potential legal models for activities in the exploration, exploitation, and utilization of space resources" and proposals for a dedicated international conference on space resources under the auspices of the United Nations".[28]

During the session, the relevant working group held nine formal and informal meetings and considered the following items:

(a) The name of the working group;
(b) The 5-year work plan and methods of work of the working group;
(c) Possible topics and areas of contributions that States members of the Committee may consider when providing information to the working group;
(d) Guidance from States members of the Committee on the international conference on space resources.[29]

4.6.2 *Defining the Name of the Working Group*

As for the name of the Working Group, and after discussion, the "Working Group on Legal Aspects of Space Resource Activities" was chosen, based on proposals from Canada and China. It should be noted that several States (Argentina, Germany, Japan, Greece) had previously supported a short version of the name (Working Group on Space Resources). The 5-year Work plan of the Working Group was

---

[26] Finland's Statement in the 61st session of the Legal Subcommittee of the UN Committee on the Peaceful Uses of Outer Space, Vienna, 28 March – 8 April 2022, Agenda Item 15 - General exchange of views on potential legal models for activities in exploration, exploitation and utilization of space resources, https://www.unoosa.org/oosa/en/ourwork/copuos/lsc/2022/statements.html (last accessed: 21.4.2023).

[27] Committee on the Peaceful Uses of Outer Space, Legal Subcommittee, Fifty-ninth session, Building blocks for the development of an international framework on space resource activities, Working paper submitted by Luxembourg and the Netherlands, A/AC.105/C.2/L.315, 3 February 2020. For the Hague Building Blocks, see the relevant developments, *infra*.

[28] Committee on the Peaceful Uses of Outer Space, Legal Subcommittee, Sixty-first session. A/AC.105/C.2/2022/CRP.21, 1 April 2022.

[29] Annex II, Report of the Chair and Vice-Chair of the Working Group on Legal Aspects of Space Resource Activities, A/AC.105/1260, para. 2.

finally adopted, after consideration, by the States, of three successive draft proposals.[30] Regarding a "dedicated international conference", the Chair and Vice-Chair undertook to "circulate a request in the intersessional period for information from States members… on issues related to and arising from its mandate, including on potential topics and issues to be addressed through a dedicated international conference convened under the auspices of the United Nations and open to Governments, invited academic and other stakeholders, subject to the availability of extrabudgetary resources".[31] Finally, a document was prepared by the Chair and Vice-Chair on "Possible Areas for Member State Input/Contributions".[32] Pursuant this document, States Members were "invited and encouraged" to provide views on the following:

- The type of space resources that fall within the mandate and scope of the Working Group.
- The type of activities that fall within the mandate and scope of the Working Group.
- The type of information to be collected by the Working Group in accordance with its mandate.
- The existing legal framework for space resource activities.
- The current practices and challenges in the implementation of the existing legal framework for such activities.
- The benefits and challenges of development of a framework for such activities.
- The relevant factors for the development of a set of initial recommended principles for such activities.
- The format, agenda, topics and other details of the dedicated conference (currently) scheduled for 2024.
- Any other background or information paper, or any other views, that Member States may wish to share.

4.7 62nd Session of the LSC (20–31 March 2023): Inputs/Views by the States Members

On March 2023, the LSC made public the inputs received by States participating in the Working Group. These inputs could serve as the starting point for the Working Group to identify the view of States regarding what falls within its mandate and the legal aspects of space resource activities.

---

[30] See https://www.unoosa.org/oosa/en/ourwork/copuos/lsc/space-resources/index.html (last visit: 22.4.2023). For the adopted work plan, see Report of the Legal Subcommittee at it sixty-first session in 2022, A/AC.105/1260, Annex II, Appendix.

[31] Report of the Chair and Vice-Chair of the Working Group on Legal Aspects of Space Resource Activities, *op. cit.*, para.7.

[32] Possible Areas for Member State Input/Contributions Circulated to States Members of the Committee by the Chair and Vice-Chair of the Working Group on Legal Aspects of Space Resource Activities, https://www.unoosa.org/documents/pdf/copuos/lsc/space-resources/Possible_Areas_for_Member_State_Input_as_at_2_April_SF.pdf (last accessed: 22.4.2023).

4.7.1 The Type of Space Resources That Fall Within the Mandate and Scope of the Working Group

Regarding the type of space resources that fall within the Working Group's mandate, States expressed different positions. Japan supported that all natural resources like minerals and water that exist in outer space should be considered.[33] The United Kingdom emphasised that only materials found in outer space fall within the mandate.[34] The same view was also expressed by France, which defined space resources as materials in the subsoil of celestial bodies. It also suggested the consideration of "*certain local in situ positions of the Moon and other celestial bodies (such as peaks of eternal light on the moon which, given their limited nature, appear to be a space resource of particular interest)*".[35] The United States proposed a definition that already existed in their national legislation, according to which a space resource is "*an abiotic resource in situ in outer space*",[36] including water and minerals. USA further referred to the "similar" definition included in the 2019 Building Blocks for the Development of an International Framework on Space Resource Activities ("*an extractable or recoverable abiotic resource in situ in outer space*"), while arguing that the mandate of the Working Group did not include "*discussion of orbital slots or radio spectrum allotment*". The latter definition was supported by Australia,[37] New Zealand,[38] Canada,[39] Luxembourg,[40] and Germany, which also suggested the consideration of topographical features of the moon and other celestial bodies and that the discussion should be structured "*according to the different types of resources, e.g. limited raw materials, essential resources to sustain human*

[33] COPUOS, LSC, 'Japan – Input to the Working Group on Legal Aspects of Space Resource Activities' (20 March 2023), UN Doc. A.AC.105/C.2/2023/CRP.33, para. 1 (hereinafter: "Japan").

[34] COPUOS, LSC, 'United Kingdom – Input to the Working Group on Legal Aspects of Space Resource Activities' (20 March 2023), UN Doc. A.AC.105/C.2/2023/CRP.21, p. 1, para. 3 (hereinafter: "UK").

[35] COPUOS, LSC, 'France – Input to the Working Group on Legal Aspects of Space Resource Activities' (20 March 2023), UN Doc. A.AC.105/C.2/2023/CRP.12, p. 1, para. A.1 (hereinafter: "France").

[36] COPUOS, LSC, 'United States – Input to the Working Group on Legal Aspects of Space Resource Activities' (20 March 2023), UN Doc. A.AC.105/C.2/2023/CRP.37, para. II (hereinafter: "USA").

[37] COPUOS, LSC, 'Australia – Input to the Working Group on Legal Aspects of Space Resource Activities' (20 March 2023), UN Doc. A.AC.105/C.2/2023/CRP.7, p. 1 (hereinafter: "Australia").

[38] COPUOS, LSC, 'New Zealand – Input to the Working Group on Legal Aspects of Space Resource Activities' (20 March 2023), UN Doc. A.AC.105/C.2/2023/CRP.18, p. 1 (hereinafter: "New Zealand").

[39] COPUOS, LSC, 'Canada – Input to the Working Group on Legal Aspects of Space Resource Activities' (20 March 2023), UN Doc. A.AC.105/C.2/2023/CRP.11, p. 1 (hereinafter: "Canada").

[40] COPUOS, LSC, 'Luxembourg – Input to the Working Group on Legal Aspects of Space Resource Activities' (20 March 2023), UN Doc. A.AC.105/C.2/2023/CRP.16, p. 5, para. 2 (hereinafter: "Luxembourg").

presence and or according to the different types of activities".[41] It is worth noting that New Zealand, the United States, France, Greece, Luxembourg, Norway[42] and the United Kingdom reaffirmed ITU's role and competence in managing radio frequencies and slots in Geostationary Orbit. Morocco[43] and Russia,[44] on the other hand, proposed broader definitions to include all types of resources, including *inter alia* orbital and frequency resources. Russia further suggested "[systematising] *space resources according to specific classification attributes*" and then working on certain terms and definitions necessary for the regulation of space resource activities.

4.7.2 The Type of Activities That Fall Within the Mandate and Scope of the Working Group

Addressing the issue of what activities fall within the working group's mandate, States expressed various positions, some more 'strict and precise' and some more 'broad and vague'. The United States proposed those activities conducted in outer space for the purpose of extraction or utilisation of those resources.[45] Russia suggested that all activities relating to the exploration, exploitation, and utilisation of space resources should be included.[46] Similarly, Norway's position also highlighted the issue of ownership, of the *in situ* use and the transfer of space resources back to Earth,[47] whereas Germany focused on the activities that have the potential to irreversibly alter the environment of outer space and exceed the scientific needs.[48] Greece argued that "*given that issues related to orbits are addressed, directly or indirectly, in other agenda items of the Committee, ... the Working Group... should focus on the in situ resource utilization, as well as the extraction and commercial exploitation of natural space resources of the celestial bodies*". However, Greece clarified that "*these two activities should not necessarily be approached together, as*

[41] COPUOS, LSC, 'Germany – Input to the Working Group on Legal Aspects of Space Resource Activities' (20 March 2023), UN Doc. A.AC.105/C.2/2023/CRP.13, para. I (hereinafter: "Germany").

[42] COPUOS, LSC, 'Norway – Input to the Working Group on Legal Aspects of Space Resource Activities' (20 March 2023), UN Doc. A.AC.105/C.2/2023/CRP.19, p. 1 (hereinafter: "Norway").

[43] COPUOS, LSC, 'Morocco – Input to the Working Group on Legal Aspects of Space Resource Activities' (20 March 2023), UN Doc. A.AC.105/C.2/2023/CRP.17, p. 2, para. 1 (hereinafter: "Morocco").

[44] COPUOS, LSC, 'Russian Federation – Input to the Working Group on Legal Aspects of Space Resource Activities' (20 March 2023), UN Doc. A.AC.105/C.2/2023/CRP.20, pp. 1–2 (hereinafter: "Russia").

[45] USA, para. II.

[46] Russia, p. 2.

[47] Norway.

[48] Germany, para. II.

they have clearly different characteristics".[49] Australia[50] and New Zealand[51] referred to the definition of space resource activity given by the Hague Working Group. Analogous was the definition put forward by Japan[52] and France, which also proposed categorising space resource activities as scientific, supportive to exploration missions and commercial.[53] The United Kingdom prioritised the examination of 'scientific' over 'for-profit' space resource activities while also including the processing of resources in space and the produced products.[54] Canada mentioned that "*the entire spectrum of planned and future activities be considered*", given that "*any rules imposed on activities around SRU may have unintended consequences on other exploration activities*".[55]

4.7.3 The Type of Information to be Collected by the Working Group in Accordance with its Mandate

Concerning the type of information that the WG should collect, a number of States, namely Japan, Australia, New Zealand, Canada, Germany and France, agreed that it should include plans for current and/or future space resource missions conducted either by governmental or non-governmental entities, related technological and scientific facts. Some States proposed that the WG should collect 'legal' information regarding national legislation,[56] the applicability of the five UN Treaties,[57] as well as other international governance regimes, like the Seabed Authority,[58] the regime of ITU and that of Antarctica.[59] Moreover, the work of the Hague Working Group and the Artemis Accords could be of guidance to the WG, which should pay attention to the views of relevant stakeholders, civil society and academia.[60]

[49] COPUOS, LSC, 'Greece – Input to the Working Group on Legal Aspects of Space Resource Activities' (20 March 2023), UN Doc. A.AC.105/C.2/2023/CRP.13, p. 3 (hereinafter: "Greece").

[50] Australia, p.1.

[51] New Zealand, p. 1.

[52] Japan, para. 2.

[53] France, para. A.2.

[54] UK, para. 4.

[55] Canada, p. 2.

[56] Japan, para. 3; Germany, para. III; France, para. 3.

[57] Australia, p.1.

[58] Norway.

[59] France, para. A.3.

[60] Germany, para. III.

4.7.4 The Existing Legal Framework for Space Resource Activities

States expressed different views regarding the existing legal framework for space resources. As Norway stated, "*there is no consensus regarding the interpretation of the international legal framework pertaining to space resource activities*".[61] Japan asserted that the OST lacks provisions regulating space resources activities.[62] On the contrary, Australia and the USA expressed the view that ownership over space resources is not prohibited under the OST, but the content of the deriving rights and obligations must be further examined.[63] Luxembourg submitted that the UN space treaties have not been tested with respect to rights over space resources and that Article II of the OST does not cover space resources.[64] Both Canada and New Zealand suggested that the existing legal framework can serve as the basis for regulating space resource activities.[65] To this end, the USA, Australia, New Zealand, and Canada suggested that the principles contained in the Artemis Accords could be enlightening.

For their part, the United Kingdom and Germany stressed the distinction between scientific and commercial activities. UK acknowledged the absence of an explicit prohibition on space resource utilisation (SRU) under the OST and envisioned collaboration between States and private entities for scientific exploration.[66] Germany shared this perspective on scientific activities and emphasised the need for a consensus-adopted investment-friendly regime for commercial activities.[67] Both Australia and France proclaimed the extension of principles contained in the Rescue and Return Agreement, the Liability Convention, the Registration Convention and the Moon Agreement to space resource activities.[68] Russia and Greece expressed the view that in regulating space resource activities, experience can be drawn from related legal frameworks such as the ITU frequency spectrum management regime, the ICAO framework for air traffic management, the administration of international seabed by the Seabed Authority and the Antarctica regime.[69] Furthermore, Russia, after having pointed out that "*the 1967 Outer Space Treaty does not restrict the use of natural resources, including their extraction. Thus, the exploitation of natural resources… is a type of outer space 'utilization', provided for by the principle of freedom of exploration and utilization of outer space. However, such utilization cannot constitute 'national appropriation' generating a right of ownership*", noted that

[61] Norway.

[62] Japan, para. 4.

[63] Australia, p. 2; USA, para. I.

[64] Luxembourg, para. 3.1.

[65] Canada, p. 2; New Zealand.

[66] UK, para. 5.

[67] Germany, para. IV.

[68] France, para. B.1.

[69] Russia, pp. 4–4; Greece, p. 3.

"*…the question is how Article II of the 1967 Outer Space Treaty should be applied and interpreted in order to provide an appropriate legal regime that would facilitate the exploration and exploitation of natural space resources, including by non-governmental entities*".[70] Finally, Belgium argued that the Moon Agreement, the only international legal regime allowing the exploitation of space resources for all nations in compliance with the OST, can be of guidance. It considered that a 'club-based' model, like Antarctica's, can be taken into account, although it does not comply with the OST principles.[71]

The United States, Japan and Luxembourg submitted to the WG their national legislation regarding space resource activities by private entities.

4.7.5 The Current Practices and Challenges in the Implementation of the Existing Legal Framework for Such Activities

Different views were expressed regarding the current practices and challenges in the implementation of the existing legal framework for such activities. The United States has already conducted or plans to conduct activities regarding the collection and return to Earth of samples from celestial bodies, both for scientific and commercial purposes.[72] Japan issued its first licence under its domestic law in 2022.[73] Canada is planning a manned mission to the Moon, although acknowledging that a legal framework regulating space resource activities is currently lacking.[74] On the contrary, Russia submitted that national laws permitting ownership over space resources are not in accordance with the non-appropriation principle and are obscuring the implementation of the OST.[75] Moreover, it proposed that the WG must distinguish between 'use' and 'appropriation by use', work on a benefit-sharing mechanism, address the issue of limited space resources in terms of volume or location and prioritisation of missions to this respect, examine the question whether the principle of free access to all areas of celestial bodies applies to infrastructures built on the territory of celestial bodies, and consider, within the existing legal framework, the issue of placement and use of "*multi-satellite systems of small spacecraft*", which, *inter alia*, increase "*the threat of additional 'space debris' generation*", thus impairing the free access to outer space and the use of

[70] Russia, p. 4.

[71] COPUOS, LSC, 'Belgium – Input to the Working Group on Legal Aspects of Space Resource Activities' (20 March 2023), UN Doc. A.AC.105/C.2/2023/CRP.36, paras. 5–6 (hereinafter: "Belgium").

[72] USA, para. I.

[73] Japan, para. 5.

[74] Canada, p. 2.

[75] Russia, pp. 5–6.

orbital-frequency resource. Australia raised the issue of registration of objects constructed on and around the Moon, especially when these objects include component parts launched from Earth by more than one State.[76] France expressed its concern that the existing legal framework may show its limits in the context of the multiplication of space activities and technological development.[77] New Zealand[78] and Japan[79] highlighted the issue of the safety and sustainability of space resource activities, similar to Luxembourg, which also called for legal certainty.[80] Finally, Morocco expressed its support for current initiatives while stressing the interests of developing States.[81]

4.7.6 The Benefits and Challenges of the Development of a Framework for Such Activities

This is where the views of the States converged. Morocco recognised that a new legal framework would be beneficial by guaranteeing equitable access to space resources.[82] Japan acknowledged the benefit of a new framework promoting safe and sustainable space resource activities while being adaptable to new technological and scientific developments.[83] Likewise, Australia and New Zealand adopted similar positions and suggested that a new set of rules would provide legal clarity and certainty to private actors in the space industry.[84] Analogous were the views expressed by Greece, Germany, Belgium and France. The United Kingdom and Canada both agreed with the aforementioned positions. However, they expressed concern that gathering the necessary information in order to understand the types of missions being proposed by States or private actors posed serious challenges.[85] Finally, Bahrain considered that a framework based on best practices would be beneficial, although meeting all States' expectations would be challenging.[86]

---

[76] Australia, p. 7.

[77] France, para. B.2.

[78] New Zealand, p. 2.

[79] Japan, para. 5.

[80] Luxembourg, p. 6.

[81] Morocco, para. 5.

[82] Morocco, para. 6.

[83] Japan, para, 6.

[84] Australia, p. 7; New Zealand, p. 3.

[85] UK, para. 6; Canada, p. 4.

[86] COPUOS, LSC, 'Bahrain – Input to the Working Group on Legal Aspects of Space Resource Activities' (20 March 2023), UN Doc. A.AC.105/C.2/2023/CRP.9, p. 2 (hereinafter: "Bahrain").

4.7.7 The Relevant Factors for the Development of a Set of Initial Recommended Principles for Such Activities

States participating in the discussions offered diverse perspectives on the factors crucial for developing initial principles for space resource activities. Japan highlighted some key points, including the definition of space resources and relevant activities, ownership, information sharing, international coordination, and dispute settlement.[87] Australia mentioned that "*general themes within relevant UN treaties and non-binding instruments can support considerations regarding the development of a set of initial recommended principles*".[88] It is worth noting that Australia, Austria and the Netherlands submitted a joint proposal providing "*considerations for developing a set of initial recommended principles for the exploration, exploitation and utilisation of space resources*", based on "*a survey of key themes which are addressed in selected instruments of the existing framework*" (the Moon Agreement, the Artemis Accords and the Hague Building Blocks).[89] New Zealand proposed the discussion on the principles developed by the Hague Working Group, informed by scientific and technical input, including planetary protection protocols.[90] The United Kingdom also mentioned planetary protection.[91] Canada emphasised that impairment to innovation in outer space activities must be avoided.[92] Morocco advocated reinforcing existing institutions, promoting cooperation, and regulating commercial space mineral use.[93] Bahrain suggested considering implementation costs and time.[94] The United States seeks an outcome in conformity with the OST, based on technical and scientific input from experts and stakeholders' involvement, which will promote predictability and legal certainty.[95] Germany emphasised that defining the will of the States is crucial for developing a set of principles.[96] Luxembourg underscored the need for a new, adaptable legal framework to accommodate emerging space actors, environmental concerns, assistance to developing states, individual rights, and the establishment of 'safety zones'.[97] Finally, Norway called for a

[87] Japan, para. 7.

[88] Australia, p. 8.

[89] COPUOS, LSC, 'Relevant considerations for developing a set of initial recommended principles foe the exploration, exploitation and utilisation of space resources' (20 March 2023), UN Doc. A.AC.105/C.2/2023/CRP.6

[90] New Zealand, p. 3.

[91] UK, para. 6.

[92] Canada, p. 4.

[93] Morocco, para. 7.

[94] Bahrain.

[95] USA, para. IV.

[96] Germany, para. VII.

[97] Luxembourg, para. 4.

framework promoting predictable and transparent conditions for safe and sustainable space resource activities based on a common understanding.[98]

4.7.8 The Format, Agenda, Topics and Other Details of the 2024 Dedicated Conference

Various positions were put forward regarding the format, topics and other details of the dedicated conference currently scheduled for 2024. Canada, Germany, Luxembourg, Germany, and Norway called for broad participation from States, academia, the commercial sector, and civil society in the conference. The United States, Australia, New Zealand, and Canada proposed an exchange of information regarding current practices relating to space resource activities and planned missions, especially regarding private actors. Belgium, Germany and Greece focused, *inter alia*, on the environmental aspects of space resource activities, while the first two also suggested the establishment of requirements for the authorisation of space resource activities.[99] The issue of space debris was put forward by Bahrain, along with the issue of access to space by emerging nations. Finally, Belgium proposed the consideration of the establishment of a dedicated authority with executive powers to promote benefit-sharing and the sustainability of space resource activities. The idea was also suggested by Russia, which favoured the creation of a monitoring mechanism or a specialised international body entrusted with issues of compliance, responsibility and dispute settlement.[100] Bahrain also acknowledged the need for a specialised court.[101]

* * *

At the 66th session of the COPUOS (31 May–9 June 2023), the Working Group agreed to five topics within the scope of which the international conference would be held, namely:

(a) Implications of the legal framework for space resource activities;
(b) The role of information-sharing in supporting space resource activities;
(c) The scope of future space resource activities;
(d) Environmental and socioeconomic aspects of space resource activities;
(e) International cooperation in scientific research and technological development for space resource activities.[102]

---

98 Norway.

99 Greece, p. 6; Belgium, Annex; Germany, para. VIII.

100 Russia, pp. 2–3.

101 Bahrain, p. 2.

102 Report of the Committee on the Peaceful Uses of Outer Space, Sixty-sixth session (31 May–9 June 2023), A/78/20, para. 234.

4.8 63rd Session of the LSC (15–26 April 2024): International Conference on Space Resources

Said conference indeed took place on April 15th, 2024, at the Vienna International Center, during the 63rd session of the LSC (15–26 April 2024). The speakers were the following: Gao Guozhu (China), Sergio Marchisio (Italy), Christopher Johnson (United States), El Bachir Zantou (Morocco), Rosa Ma. Ramírez de Arellano y Haro (Mexico). Rocío Germania Narváez Benalcázar (Ecuador), Georgios Kyriakopoulos (Greece). Marcus Schladebach (Germany), Vasiliy Gudnov (Russian Federation), Su Jinyuan (China), Tanja Masson-Zwaan (Netherlands), Konrad Marciniak (Poland), Yury Razoumny (Russian Federation), Ian Grosner (Brazil), Mathias Link (Luxembourg), and Ryosuke Murayama (Japan). Before the conference, an event was held in Luxembourg, during the Space Resources Week, entitled "Expert meeting collecting preliminary inputs for consideration at the international conference in Vienna in 2024". Said event was co-hosted by Belgium and Luxembourg and organized in cooperation with the United Nations.[103]

One thing that was clear from the contributions in the conference was that the use of space resources is now beyond question. All the debate and controversy among States is about defining the conditions under which this will take place, as, at present, there is no specific international legal regime for the use and exploitation of space resources. Further, it seems that States already agree on a minimum set of requirements in this area, such as the need for sustainable use of resources, the exchange of information, and the need to avoid friction and conflict. This is undoubtedly a good starting point for the international community to move towards more detailed rules on the use and exploitation of space resources.

At the same session, the Chair and Vice-Chair of the Working Group invited States members and permanent observers to provide contributions in relation to an initial set of recommended principles for space resource activities, according to the work plan of the Working Group, with a deadline for contributions of 16 September 2024.[104] In this respect, the Working Group "noted a number of elements…including, but not limited to, the centrality of the provisions of the Outer Space Treaty; the principles of non-appropriation and the peaceful use of outer space; matters related to the nature and purpose of space resources and their potential benefits for both scientific research and economic development; the economic, environmental and equitable aspects of space resource activities; the protection of interests of current and future space actors; ethical, indigenous and intergenerational aspects; international coordination and consultation measures; licensing authority mechanisms; private sector involvement; aspects related to investments and international trade; aspects of benefit-sharing and interests of developing countries".[105] Based on the

---

[103] *Idem*, para. 232.

[104] Report of the Chair and Vice-Chair of the Working Group on Legal Aspects of Space Resource Activities, 23 April 2024, A/AC.105/C.2/2024/SRA/L.1, para. 8.

[105] *Idem*, para. 7.

discussions at the session as well as on the contributions received, the Chair and the Vice-Chair of the Working Group would prepare an initial draft of recommended principles for space resource activities.[106]

***

The discussions and negotiations on the issue of space resources, as detailed in this Part, demonstrate the international community's strong and growing interest in this subject. It is no coincidence that, since its inclusion on the agenda of the Legal Subcommittee (LSC), this topic has virtually dominated the discussions, also occupying a significant portion of the Committee's time and attention.

Several noteworthy conclusions can be drawn from these debates. First, the relevant international dialogue remains in its early stages. No issue has been definitively resolved. Even the specific resources subject to regulation are still under definition, and the legal regime to govern them is far from being fully developed.

Broadly speaking, although the lines between them remain blurred, certain groupings of states with shared aspirations appear to be taking shape. At this stage, two main trends are emerging. On one hand, there is the view that the exploration and exploitation of space resources require a dedicated international legal regime. However, it remains unclear whether, and to what extent, such a regime would involve a departure from the fundamental principles of the Outer Space Treaty (OST)—a notion that, while not openly acknowledged, appears to be subtly gaining ground. In this context, the objectives outlined in Article 11 para. 7, of the Moon Agreement are largely respected in principle, despite this agreement's limited acceptance.

On the other hand, a second approach argues that the existing legal framework of the OST is fundamentally sufficient. According to this view, the issue is not purely legal but has strong political and economic dimensions that must be taken into account towards a system of governance of space resources, which will take into account the interests of the private sector and will include, for regulatory purposes, the establishment of rules at the national level.

Nevertheless, these trends remain fluid and not yet fully consolidated; positions may shift as the discussions evolve. It is also worth noting the relative silence of most developing countries on this issue—at least thus far—which stands in contrast to their active engagement during the negotiations of Part XI of the UNCLOS (concerning international seabed resources) during the Third United Nations Conference on the Law of the Sea. The eventual regrouping of perspectives and formal articulation of state positions at a more advanced stage of negotiations will undoubtedly be influenced, inter alia, by the growing number of States becoming parties to the Artemis Accords, as will be discussed below.

Taking the above into account, the establishment of the Working Group on Outer Space Resources is a particularly positive step towards enriching and revitalizing the dialogue between States. If allowed to operate beyond narrow political

---

[106] *Idem*, para. 9.

considerations, the Working Group could prove to be the matrix for substantive proposals and, ultimately, solutions to the thorny issue of the utilization and exploitation of space resources. The success of this effort will hinge on the integrity and relevance of an initial set of recommended principles currently being developed within the Working Group. The commitment of States—especially space-faring ones—will be crucial in determining whether these emerging principles offer meaningful guidance in the debate over the use and exploitation of space resources, or whether they will ultimately become yet another symbolic set of principles lacking regulatory force.

Chapter 5
Uncertainty of the Next Step #3: Exploring Appropriate Responses Through International Instruments and Projects

Beyond the academic discussions and the diplomatic interactions among States within the COPUOS, a range of other initiatives has emerged at the international level, aiming to establish a coherent framework for the regulation of space resource exploration and utilization. The most prominent among these—whether specifically focused on the issue at hand, such as the Hague Building Blocks, or integrated within broader national and multilateral programs for lunar return, such as the Artemis Accords and the International Lunar Research Station—are presented below.

5.1 The Hague International Space Resources Governance Working Group

Apart from the discussions taking place within UN COPUOS, the Hague Working Group on International Governance of Space Resources was another *forum* that addressed the issue of space mineral resources. The Hague Group was established in 2016, after a roundtable discussion on the Governance of Space Resources that was organized by the Hague Institute for Global Justice on 1 December 2014,[1] to assess the need for a governance framework for space resources and was composed of members and observers representing governments as well as business, academia, international organizations, NGOs and civil society. The mandate of the Group was to develop a set of "Building blocks" for the governance of space resource activities

[1] See the relevant info in: Committee on the Peaceful Uses of Outer Space, Legal Subcommittee, Fifty-seventh session, Vienna, 9–20 April 2018, The Hague Space Resources Governance Working Group l Information provided by the Netherlands, A/AC.105/C.2/2018/CRP.18, 12 April 2018, p. 1.

G. D. Kyriakopoulos, *Space Resource Utilization and Exploitation: The Suspended Step of International Law*, Space and Society,
https://doi.org/10.1007/978-3-031-96336-0_5

as a basis for "negotiations on an international agreement or non-legally binding instrument".[2]

Draft Building Blocks were presented on 13 September 2017, whereas the Working Group adopted the final text of the Building Blocks on 12 November 2019.[3] According to the proposed blocks, the purpose of an international framework ("consistent with international law") should be the creation "of an enabling environment for space resource activities" that takes into account all interests and benefits all countries and humankind. In this respect, the international framework should include relevant recommendations and promote the identification of 'best practices' by States, international organizations and non-governmental entities. It should be noted that the building blocks recognize the potential implications of the issue of the use of space resources for the maintenance of international security: According to the proposed Principles, the future international framework "should be designed to: ... prevent disputes arising out of space resources [and] should provide that ...space resources shall be used exclusively for peaceful purposes".[4]

5.1.1 "Space Resource", "Utilization of Space Resources"

In the context of the building blocks, "space resource" is defined as "an extractable and/or recoverable[5] abiotic resource in situ in outer space"; it is further clarified that "according to the understanding of the Working Group, this includes mineral and volatile materials, including water, but excludes (a) satellite orbits; (b) radio spectrum; and (c) energy from the sun except when collected from unique and scarce locations".[6] Unique and scarce locations include "peaks of eternal lights", which are highland regions that receive continuous sunlight, thereby having the capacity to serve as a permanent power source.[7] A restrictive approach to the concept of space resources is thus adopted, essentially referring to the natural resources of celestial

[2] See https://www.universiteitleiden.nl/en/law/institute-of-public-law/institute-of-air-space-law/the-hague-space-resources-governance-working-group (last accessed: 23.04.2023).

[3] Building Blocks for the Development of an International Framework on Space Resource Activities, November 2019 (hereinafter: "Building Blocks"). Text *in:* https://www.universiteitleiden.nl/binaries/content/assets/rechtsgeleerdheid/instituut-voor-publiekrecht/lucht%2D%2Denruimterecht/space-resources/bb-thissrwg%2D%2Dcover.pdf (last accessed: 10.07.2024).

[4] Building Blocks, 4.2 (d), 4.3 (a).

[5] In view of advancing technology, recoverability was added to the definition along with extractability to consider the possibility for space resources to be harvested or collected from the surface of a celestial body, see O. de O. Bittencourt Neto, M. Hofmann, T. Masson-Zwaan & D. Stefoudi (Eds.), Building Blocks for the Development of an International Framework for the Governance of Space Resource Activities: A Commentary, Eleven International Publishing 2020, p. 25 (hereinafter: "Building Blocks Commentary").

[6] Building Blocks, 2.1.

[7] Building Blocks Commentary, p. 23, Footnote 8.

bodies within the solar system.[8] This means that the Hague Building Blocks refer to the resources of planets, asteroids and comets within the solar system, i.e. without adopting the view, expressed by some scholars, of a potential distinction between planets (considered as the only "celestial bodies") and other "bodies" (asteroids, comets, etc.).[9] The methods of acquisition of space resources are not further specified in this context.[10] Further, "utilization of space resources" is considered to be "the recovery of space resources and the extraction of raw mineral or volatile materials therefrom",[11] which would mainly suggest "space mining" activities. It is specified, in this respect, that "according to the understanding of the Working Group, this excludes secondary utilization of space resources, i.e. (a) utilization of raw materials derived from space resources; and (b) marketing and distribution of space resources". Following the principle of adaptive governance,[12] the Working Group chose to implement this exclusion to set a limit, or "cut-off point", for applicability, taking into account the characteristics of the Building Blocks.[13] However, if secondary utilization were to be excluded, the question arises as to what the purpose of the extraction itself is. It is also remains unclear, what the extraction and collection of space resources would mean in the context of ownership rights for private companies. Does such a right imply a permission for space-mining corporations to engage in exhaustive mining of celestial bodies until they have been entirely depleted and what does this mean in relation to the prohibition of national appropriation?

5.1.2 The "Principles"

The international regime (which, it should be noted, is not identified as "legal") to which the building blocks refer is governed by a number of principles, which are as follows:

4. Principles

   4.1 The international framework should be consistent with international law.
   4.2 The international framework should be designed to:

      a) Adhere to the principle of adaptive governance by incrementally regulating space resource activities at the appropriate time;

---

[8] Building Blocks, 3.2.

[9] See *supra*, under Part II.

[10] Building Blocks Commentary, p. 23.

[11] Building Blocks, 2.2.

[12] The concept of adaptive governance advocates for the incremental and timely regulation of space resource activities and played a pivotal role in the drafting of the Building Blocks, see Building Blocks Commentary, p. 32.

[13] Building Blocks Commentary, p. 24.

b) Promote compatibility and predictability of domestic frameworks of States and internal frameworks of international organizations;
c) Contribute to sustainable development;
d) Prevent disputes arising out of space resource activities;
e) Promote and secure the orderly and safe utilization of space resources;
f) Promote the sustainable, rational, efficient and economic use of space resources;
g) Promote the use of sustainable technology;
h) Provide legal certainty and predictability for operators;
i) Take into particular account the needs of developing countries;
j) Take into particular account the needs of science;
k) Take into particular account the contributions of pioneer operators.

4.3 The international framework should provide that:

a) Space resources shall be used exclusively for peaceful purposes;
b) Space resource activities shall be carried out for the benefit and in the interests of all countries and humankind irrespective of their degree of economic and scientific development;
c) Appropriate international consultations shall be undertaken in accordance with Article IX OST if there is a reason to believe that any potentially harmful interference may be caused;
d) International cooperation in space resource activities shall be conducted in accordance with international law.

It should first be noted that the (proposed) framework is simply required to be "consistent with international law". The use of this language can be interpreted as if the drafters of the principles wished them to have a general orientation towards international law, without requiring full compliance with it, as would obviously be the case if the phrase 'in accordance with international law' had been used instead of simply 'consistent with international law'. A view that is shared by the authors of the Building Blocks commentary, since applicable provisions of international law ought to only be "observed".[14] This differentiation is also reinforced by the fact that, in other parts of the 'principles', the phrase 'in accordance' has been used [4.3, (c) and (d)]. In this connection, it is to be recalled that Article III OST requires that "States Parties to the Treaty shall carry on activities in the exploration and use of outer space, including the Moon and other celestial bodies, *in accordance with international law…*", while Article I of the same treaty also provides that "Outer space, including the Moon and other celestial bodies, shall be free for exploration and use by all States without discrimination of any kind, on a basis of equality and *in accordance with international law*" (emphasis added). Principle 4.2 underlines the relevance of the principle of adaptive governance and the promotion of the harmonious co-existence of different domestic and internal frameworks of international

[14] Building Blocks Commentary, p. 32.

organizations in line with the Building Blocks.[15] This principle serves the purpose of avoiding superfluous conflict.[16]

Finally, it is noted that, according to Principle 4.2 of the Hague Group's building blocks, "the international framework should be designed to: ...f) promote the *sustainable, rational, efficient and economic* use of space resources" (emphasis added). This language is similar to that already used in the context of Article 44(1)(b) of the Constitution of the International Telecommunication Union (ITU), which reads as follows:

> CONSTITUTION OF THE INTERNATIONAL TELECOMMUNICATION UNION
> ARTICLE 44
> Use of the Radio-Frequency Spectrum and of the Geostationary-Satellite and Other Satellite Orbits
> ...
> 2. In using frequency bands for radio services, Member States shall bear in mind that radio frequencies and any associated orbits, including the geostationary-satellite orbit, are limited natural resources and that *they must be used rationally, efficiently and economically*, in conformity with the provisions of the Radio Regulations, *so that countries or groups of countries may have equitable access to those orbits and frequencies, taking into account the special needs of the developing countries and the geographical situation of particular countries* (emphasis added).

The comment that the wording of Principle 4.2 was inspired by the ITU Constitution was also made by the authors of the commentary to the Building Blocks.[17] The—essential—difference between the two texts lies in the absence of a reference to "equitable access" for all States in Principle 4.2(f) of the Building Blocks—a provision according to which, at the ITU level, all States are entitled to radio frequencies—leaving open, in the context of the Building Blocks, the prospect of an exploitation of the resources of the celestial bodies according to the principle of "first come, first served".[18] It is noted that radio frequencies (as well as orbits) are considered to be limited natural resources, in accordance with Article 44 of the ITU Constitution. Principle 4.2(f) is indicative of the overall spirit and approach of the Building Blocks. The language used is of particular importance, especially since, according to Principle 4.2(h), "The international framework should be designed to… Provide legal certainty and predictability for operators". Principle 4.2 (h) is viewed as paramount for the space industry since it takes into account the wishes of private operators for legal clarity and predictability. Legal ambiguity regarding the requirements for mining licenses could not only impact the operators' business strategies, but also their research and development plans.[19]

---

[15] Building Blocks Commentary, p. 32, see Principle 4.2 (a) and (b).

[16] Building Blocks Commentary, p. 32, see Principle 4.2 (b).

[17] Building Blocks Commentary, p. 33.

[18] See, in this respect, F. von der Dunk (Ed.) (with F. Tronchetti), *Handbook of Space Law*, Edward Elgar, 2015, p. 480 *et seq.*

[19] Building Blocks Commentary, p. 32.

5.1.3 "Priority Rights", "Resource Rights"

Other building blocks provide for the granting of preferential rights to potential operators and resource rights "over raw mineral and volatile materials extracted from space resources":

7. Priority rights

The international framework should enable the attribution of priority rights to an operator to search for and/or recover space resources for a maximum period of time and a maximum area upon registration in an international registry, and provide for the international recognition of such priority rights. The attribution, duration, and the area of the priority right should be determined on the basis of the specific circumstances of a proposed space resource activity.

8. Resource rights

   8.1 The international framework should ensure that resource rights over raw mineral and volatile materials extracted from space resources, as well as products derived therefrom, can lawfully be acquired through domestic legislation, bilateral agreements and/or multilateral agreements.
   8.2 The international framework should enable the mutual recognition between States of such resource rights.
   8.3 The international framework should ensure that the utilization of space resources is carried out in accordance with the principle of non-appropriation under Article II OST.

The purpose of priority rights is to ensure an effective mechanism for managing space resource activities "in accordance with international law", the necessary attribution of rights to its operators to safeguard their efforts and investments in locating space resources and to coordinate access to resources to prevent harmful interference with other operators in the process.[20] These rights are time-limited and activated by registration in an international registry but grant operators precedence in the search and recovery of space resources.[21] According to Building Block 7, the international framework should guarantee universal recognition of priority rights, making an international registry a fundamental requisite for the attribution of these rights.[22] It is important to note, however, that priority rights, including their scope and duration, should be decided depending "on the particular, actual circumstances" of the space resource activity in question.[23] However, priority rights host the potential for inequitable outcomes since many developing nations do not have the financial means to engage in space operations which, as a

[20] Building Blocks Commentary, p. 46.

[21] Building Blocks Commentary, *idem*.

[22] Building Blocks Commentary, p. 47.

[23] Building Blocks Commentary, p. 47.

consequence, could lead to the exploitation (of space resources) being ensured only by the most technologically advanced space-faring nations.

This view is shared by other scholars by also emphasizing that priority rights should be subject to regulations to prevent the risk of monopoly and uphold the principle of freedom of exploration and use.[24] According to other opinions, priority rights ought to be granted solely to those companies that have a physical presence *in situ* as opposed to simply identifying space resources *via* remote sensing which would prevent the concession of claims to entities devoid of any real intent or capability of mining.[25] This, in turn, would prevent the occurrence of impediments to the efficient use of space resources.[26]

Resource rights aim to facilitate the commercialization of space resources.[27] These rights are intended to be legally acquired under the proposed international framework and can be granted through domestic legislation or international agreements.[28] Mutual recognition of resource rights between States is proposed to encourage the development of space resource activities.[29] Paragraph 8.3 again highlights that the application of this Building Block should be consistent with international law to ensure the lawful utilization of space resources.[30] (According to the Commentary of the Building Blocks) The principle of non-appropriation, already enshrined in Article II of the Outer Space Treaty, should be recognized in this context to ensure that the use of space resources is in accordance with international law.[31]

Since such rights do not find any grounds in applicable international space law, nor does the building blocks attempt any interpretation of existing principles and norms of space law that would allow such a grounding, these rights should be understood as orientations *de lege ferenda* for the formulation of a framework for the exploitation of space resources according to patterns that are both acceptable and desirable to the authors of the building blocks. It should be recalled that, with regard to Building Block 7, the commentary recognized that the determination of "the applicable duration and area of effect" of these rights should be guided by the actual circumstances of the space resource activity.[32]

---

[24] F. Xu & J. Su, "New Elements in The Hague Space Resources Governance Working Group's Building Blocks", Space Policy, Vol. 53, article id. 101386, August 2020, p. 3.

[25] Ch. Davis & M.J. Sundahl, "The Hague Working Group on Space Resources: Creating the Legal Building Blocks for a New Industry", The Air & Space Lawyer, Vol. 30, No 3, Fall 2017, p. 3.

[26] Davis & Sundahl, op. cit.

[27] Building Blocks Commentary, p. 50.

[28] Building Blocks Commentary, p. 50.

[29] *Idem.*

[30] *Ibid.*

[31] *Ibid.*

[32] Building Blocks Commentary, p. 47.

5.1.4 Establishment of Safety Zones Around Space Resource Activities

Building Block 11 proposes safety measures and acknowledges the inherent risks of space resource activities.[33] This section encourages several actions to increase transparency and international cooperation by promoting the development of procedures and conformity assessments to avoid harmful impacts.[34] In that regard, also the compliance with common technical standards as well as the standardization of equipment, procedures and processes must be ensured in order to maintain a sustainable use of outer space.[35]

Moreover, paragraphs 11.3 and 11.4 of the Building Blocks provide for the establishment of safety zones around areas identified for space resource activities:

> 11. Technical standards for, prior review of, and safety zones around space resource activities
>
> …
>
> 11.3 Taking into account the principle of non-appropriation under Article II OST, the international framework should permit States and international organizations responsible for space resource activities to establish a safety zone, or other area-based safety measure, around an area identified for a space resource activity as necessary to assure safety and to avoid any harmful interference with that space resource activity. Such safety measure shall not impede the free access, in accordance with international law, to any area of outer space by personnel, vehicles and equipment of another operator. In accordance with the area-based safety measure, a State or international organization may restrict access for a limited period of time, provided that timely public notice has been given setting out the reasons for such restriction.
>
> 11.4 The international framework should provide that appropriate international consultations are undertaken in case of possible overlap of safety zones or conflicts involving the freedom of access recognized by international law.

The main function of these safety zones is to protect space resource activities and prevent harmful interference with other space operators.[36] The establishment of these zones require a careful balancing act between safety, on the one side, and the principles of non-appropriation and freedom of access to all space areas, on the other side, as stipulated in Article II OST.

---

[33] Building Blocks Commentary, p. 65.

[34] Building Blocks Commentary, p. 66.

[35] *Idem.*

[36] *Ibid.*

While the establishment of safety zones may serve a legitimate purpose in providing a safe environment for space operators to extract and collect space resources, legal uncertainty remains with respect to the following issues:

To begin with, it is not immediately clear why the proposed establishment of safety zones is said to occur "taking into account the principle of non-appropriation under Article II OST", given that—under its prevailing interpretation—this very principle poses significant legal obstacles to the exploitation of mineral space resources. It has been advocated that the original purpose of the conclusion of the OST was to hinder the United States and the Soviet Union from extraterritorial land acquisition and not to set regulation for the extraction and collection of space resources.[37] However, as has been shown elsewhere, the principle of non-appropriation is generally accepted in its broadest sense. A more specific interpretation of it, such as that contained in the Artemis Accords, requires a wider acceptance, at a universal level, which, despite the considerable acceptance of these Accords to date, cannot yet be said to exist.[38] In contrast, Article IX OST would be more appropriate to this end, since it provides for an obligation of consultations in case a State "has reason to believe that an activity... would cause potentially harmful interference with activities of other States Parties in the peaceful exploration and use of outer space".

Furthermore, the same comment about *de lege ferenda* considerations also applies to paragraph 11.3, since the creation, unilaterally by a State (as implied in paragraph 11.3), of safety zones on celestial bodies cannot be supported in the applicable international law of outer space. In particular, the combined reading of Articles I, II, III and IX OST does not, *de lege lata*, give States the right to declare unilaterally—i.e. without consultations with other States—such safety zones.[39] In this respect, an inconsistency in this paragraph is mentioned: While, at first, it is stated that "Such safety measure shall not impede the free access, in accordance with international law, to any area of outer space by personnel, vehicles and equipment of another operator" (how would it possibly not do so?), it is further stated that "In accordance with the area-based safety measure, a State or international organization may restrict access for a limited period of time, provided that timely public notice has been given setting out the reasons for such restriction". In other words, it is difficult to understand how the establishment of such zones—regardless of their stated purpose—would not interfere with the right of other States to freely access celestial bodies, as guaranteed under the Outer Space Treaty. As stated by other scholars, even though the establishment of safety zones cannot be considered as appropriation due to their time-limited access, these zones do bear an inherent risk

[37] F. Xu & J. Su, *op. cit.*, p. 3.

[38] For the Artemis Accords, see *infra*, IV.2.

[39] "...any kind of attempts to establish zones for certain periods of time with the intention to exercise jurisdiction or control over them and exclude others from using or accessing them would be contradictory to Articles I and II of the Outer Space Treaty" – S. Hobe, "The Artemis Accords: What They Mean for the Development of International Space Law", Zeitschrift fur Luft- und Weltraumrecht (German Journal of Air and Space Law), vol. 70, no. 1, 2021, p. 9.

of *de facto* appropriation if not adequately regulated.[40] Additionally, paragraph 11.3 fails to elucidate what reasons could lead to such a confinement of use of safety zones and lacks the exact requirements of what constitutes a "timely public notice" in that respect.

Finally, it would be interesting to consider the status of airspace in relation to the establishment of restriction zones or zones of special flight requirements. In airspace law, prohibited areas can only be unilaterally established above the territory of a State (land, territorial waters), in accordance with Article 9 of the 1944 Chicago Convention. So-called danger areas are further established in accordance with the provisions of Annexes 2 and 11 to the Chicago Convention and the Regional Air Navigation Plans. Restricted areas can also be established within the national airspace of States, under Annex 2. Furthermore, the establishment of the so-called "Air Defense Identification Zones" is of disputed legality, being a unilateral extension of state jurisdiction in international airspace. Finally, the only possibility of having legally accepted, restricted zones in international airspace, is the case for "no-fly zones", which are established within the Flight Information Region (FIR) of a State that was subject to UN sanctions. In this case, the Security Council shall also establish jurisdiction for the control of the no-fly zone in favour of the States which have been involved in the sanctions.[41]

It follows, therefore, that, in airspace, flight restriction areas are introduced unilaterally only within the framework of the territorial sovereignty of a State, whereas, for such restrictions in international airspace, either a decision of the UN Security Council is required or in the context of the exercise of functional jurisdiction, by a State, within a Flight Information Region, on the basis of international aeronautical regulations under the 1944 Chicago Convention and its Annexes. It is therefore difficult to justify how such restrictions could be imposed unilaterally, particularly in light of the unique legal status of outer space (*res communis omnium*).

In conclusion: The Hague building blocks introduce, in a systematic way, a model international framework for space resource activities. However, this framework can be seen, at least at present (as far as the current stage of development of space law is concerned), mainly as an interesting, *de lege ferenda* proposal to the international community, since many of its points do not seem to be reflected in international space law in force. If the international space community were ever to decide to move in such a direction in relation to the regulation of the utilization and exploitation of space resources, it is obvious that applicable space law would have to be substantially amended.

---

[40] F. Xu, J. Su & M. Mehdi, "A Re-Examination of Fundamental Principles of International Space Law at the Dawn of Space Mining", Journal of Space Law, Vol. 44, No 1, 2020, p. 26.

[41] Such no-fly zones were established by Security Council resolution in Iraq (670/25.9.90), again in Iraq north of the 36th parallel (688/5.4.91), in Libya (31.3.92 and in 2011) and in Bosnia-Herzegovina (781/12.10.92). For the establishment of restricted, *lato sensu*, zones in airspace, see the relevant developments in: A. Yokaris & G. Kyriakopoulos, International Law of Airspace, Outer Space, Nomiki Vivliothiki, 2020, p. 147 et seq. (in Greek).

5.2 The Artemis Accords

The Artemis Accords are multilateral agreements that have been proposed by the United States and already signed by other governments.[42] They are linked to the Artemis programme, a US-led initiative to return to the Moon as a first step in the further exploration of space, to Mars and beyond, to the farthest reaches of our solar system.

The Artemis accords were initially signed by the space agencies of eight countries (US, UK, UAE, Luxembourg, Japan, Italy, Canada, Australia) on 13.10.2020. As of April 8, 2025, 54 countries have joined the agreements (Angola, Argentina, Armenia, Australia, Bahrain, Bangladesh, Belgium, Brazil, Bulgaria, Canada, Chile, Colombia, Cyprus, Czech Republic, Denmark, Dominican Republic, Ecuador, Estonia, Finland, France, Germany, Greece, Iceland, India, Israel, Italy, Japan, Liechtenstein, Lithuania, Luxembourg, Mexico, Netherlands, New Zealand, Nigeria, Panama, Peru, Poland, the Republic of Korea, Romania, Rwanda, Saudi Arabia, Singapore, Slovakia, Slovenia, Spain, Sweden, Switzerland, Thailand, Ukraine, United Arab Emirates, United Kingdom, United States, and Uruguay).[43]

5.2.1 Binding Force of the Artemis Accords

The question arises as to whether the Artemis Accords constitute a treaty within the meaning of Article 2 of the 1969 Vienna Convention on the Law of Treaties (VCLT), or a non-legally binding instrument. The preamble to the Accords expresses the desire of the parties «to implement the provisions of the Outer Space Treaty and other relevant international instruments and thereby establish a political understanding regarding mutually beneficial practices for the future exploration and use of outer space, with a focus on activities conducted in support of the Artemis Program». The US State Department website also states that the Artemis Accords "…are a non-binding set of principles… [grounded in the Outer Space Treaty of 1967 (OST)] designed to guide civil space exploration and use in the 21st century". However, it is recognized that the legal and political elements co-exist in inter-state relations and disputes, and it is difficult to fully distinguish one from the other.[44] After all, it

[42] The Artemis Accords – Principles for cooperation in the civil exploration and use of the Moon, Mars, comets, and asteroids for peaceful purposes. Text in: https://www.nasa.gov/wp-content/uploads/2022/11/Artemis-Accords-signed-13Oct2020.pdf?emrc=653a00 (last accessed: 19.04.2025).

[43] See https://www.nasa.gov/artemis-accords/ (last accessed: 19.04.2025). As Hobe rightly notes, "the Artemis Accords are a unilaterally formulated text, which other Parties may sign, without having participated in the formulation of its provisions" – S. Hobe, "The Artemis Accords: What They Mean for the Development of International Space Law", Zeitschrift fur Luft- und Weltraumrecht (German Journal of Air and Space Law), vol. 70, no. 1, 2021, p. 5.

[44] See, in this respect, Emm. Roucounas, Dimossio Diethnes Dikaio (Public international law) (in Greek), Nomiki Bibliothiki, 4th edition, 2021, p. 532; ICJ, Aegean Sea Continental Shelf, Judgment, I.C.J. Reports 1978, p. 3, at par. 31.

is difficult to imagine a treaty that does not have any legal effect at all.[45] Moreover, a treaty is defined by its content and not by its designation.[46] That said, there is no explicit provision in the body of the Accords for the voluntary application of what is agreed here, which is usually technically specified in soft law texts.[47]

It is important to note that, regarding form, the Artemis Accords meet the definition of a treaty as set forth in Article 2 VCLT. This is because they constitute "an international agreement concluded between States in written form and governed by international law, whether embodied in a single instrument or in two or more related instruments and whatever its particular designation". The requisite criterion of an "agreement between States" is satisfied, as the signature was made on behalf of the States in question ("On behalf of… government"), represented, in most cases, by the heads of the space agencies, but also, in some cases, by Ministers (e.g. Italy, Japan, Luxembourg, United Arab Emirates, Lithuania, Sweden) or even by the Minister of Foreign Affairs (e.g. Greece). This practice is consistent with the provisions of Article 7 par. 1(b) VCLT on full powers («intention… to consider that person as representing the State for such purposes and to dispense with full powers»).[48]

An additional prerequisite for the formation of an international treaty (despite the absence of an explicit mention in the VCLT definition) is that the pertinent text must encompass a minimum legal content, thereby justifying the invocation of Article 26 VCLT.[49] This clearly follows from the drafting history of VCLT. According to Sir Gerald Fitzmaurice, "a treaty... is... intended to create rights and obligations,

[45] As Paul Reuter had pointed out, "un traité peut avoir d'autres effets juridiques que de créer ou modifier une « norme ». Ensuite si des effets juridiques d'un traité peuvent en effet dans certains cas être très réduits, il est difficile d'imaginer qu'un traité n'implique aucun effet juridique de quelque nature que ce soit: un traité, même en le réduisant à un simple comportement vient prendre sa place dans une série d'autres actes et comportements en relation les uns avec les autres; il définit des intérêts, justifie ou précise des interprétations, suscite des expectatives légitimes, qui se situent sur le plan du droit. Plus la forme de l'accord est solennelle plus le désir d'inscrire dans une réalité qui est à la fois « juridique » et « politique » est évident" – Paul Reuter, Introduction au droit des traités, PUF, 3e éd., 1995, Section II, §2, para. 90, online in: https://books.openedition.org/iheid/1748 (last accessed: 03.08.2024).

[46] Art. 2 par. 1 (a) VCLT; Also see: "Terminology is not a determinant factor as to the character of an international agreement or undertaking" – ICJ, South West Africa Cases (Ethiopia v. South Africa; Liberia v. South Africa), Preliminary Objections, Judgment of 21 December 1962: I.C.J. Reports 1962, p. 319, at p. 331.

[47] See, for instance, UNOOSA, Guidelines for the Long-term Sustainability of Outer Space Activities of the Committee on the Peaceful Uses of Outer Space, "The guidelines are voluntary and not legally binding under international law…" (par. 15); UNOOSA, Space Debris Mitigation Guidelines of the Committee on the Peaceful Uses of Outer Space, "These guidelines …are not legally binding under international law" (p. 2).

[48] See in this respect J. Klabbers, The Concept of Treaty in International Law, Kluwer Law International, The Hague/London/Boston, 1996, p. 31. Also see Villiger, who mentions that Article 7 VCLT "[The provision] is… sufficiently flexible to meet the needs of modern State practice" – M.E. Villiger, Commentary on the 1969 Vienna Convention on the Law of Treaties, Martinus Nijhoff, 2009, p. 146.

[49] Article 26 - Pacta sunt servanda. Every treaty in force is binding upon the parties to it and must be performed by them in good faith.

or to establish relationships, governed by international law".[50] Similarly, Sir Hersch Lauterpacht emphasized that the treaties are "agreements between States, including organizations of States, intended to create legal rights and obligations of the parties".[51] Regardless of whether the Artemis Accords constitute a political understanding, these conditions (creation of legal rights and obligations, establishment of relationships governed by international law) seem to be clearly met, insofar as they either reiterate or (and) extend principles of the four space treaties[52] or—and this is where their importance and the greatest legal interest lie—they promote a new interpretative approach to the principle of non-appropriation (Article II OST) and create a new right for the States Parties, which does not derive from existing space law, namely the right to establish safety zones,[53] which is discussed in more detail below.[54] This is supported by the emphatic insistence on the 'commitment' to the principles referred to, which could be a sign of a 'hard law' text.[55]

However, even if the Artemis Accords are indeed non-binding, the two points mentioned above (interpretation of Article II of the Outer Space Treaty (OST) and establishment of safety zones) cannot be considered to derive directly from the OST or from any other space treaty. Rather, they represent novel, original content within the Accords,[56] as an attempt to interpret the OST in a way that does not appear to be based on the letter of the Treaty or the currently accepted interpretations of critical provisions of it (such as the non-appropriation principle of Article II). Such a novel interpretation should be considered binding on the Parties. Otherwise, the relevant

---

[50] Yearbook of the International Law Commission, 1956, vol. II, p. 107.

[51] Yearbook of the International Law Commission, 1953, vol. II, p. 93.

[52] Section 3: Article IV(2) OST; Section 4: Article XI OST; Section 6: Article V OST, ARRA; Section 7: Article VIII OST, REG; Section 11: Article IX OST; Section 12: Space debris mitigation guidelines.

[53] "The provisions of safety zones are novel ones" – Y. Osada, "Governance of Space Resources Activities: In the Wake of the Artemis Accords", Georgetown Journal of International Law, vol. 53, no. 3, Spring 2022, p. 423.

[54] See Neef, which observes that "Despite being merely political agreements, the value of the Accords as a normative instrument to interpret existing law and as a potential new mode of governance is noteworthy" – R. Neef, "Artemis Accords: A New Path Forward for Space Lawmaking?", Adelaide Law Review, vol. 42, No 2, 2021, p. 576.

[55] Preamble: "COMMIT to the following principles"; Section 4: "The Signatories are committed…"; Section 6,7,8, 10, 11, 12: "The Signatories commit to…". For the importance of the language used, see O. Schachter, "The Twilight Existence of Nonbinding International Agreements", AJIL, Vol. 71, N° 2 (Apr., 1977), p. 297–298.

[56] "…this provision (Section 11) is very sensitive as it, too, introduces a term that is new to the corpus iuris spatialis: 'safety zones' " – Hobe, op. cit., p. 9. See also Osada: "It should be noted, however, that [the Accords] also contain[s[] new norms ("legal novelty") such as preservation of outer space heritage or deconfliction of activities that cannot be found in existing international space law" – Osada, op. cit., p. 411. And Deplano: "certain provisions of the Artemis Accords go a step further than mere implementation, effectively introducing concepts and principles not mentioned in the Outer Space Treaty, thus raising issues of compatibility" – R. Deplano, "The Artemis Accords: Evolution or Revolution in International Space Law?", International and Comparative Law Quarterly, vol. 70, no. 3, July 2021, p. 801–802.

sections of the Accords (10, 11) would be meaningless, given their apparent incompatibility with current international legislation.

In the light of what has been said above, the Artemis Accords, despite statements to the contrary, are international agreements in the context of which a certain degree of binding force creeps in within their general political framework. In fact, the agreements in question cannot simply be described as purely political and non-binding (for the reasons given above) and should be regarded as *sui generis* agreements.[57,58]

5.2.2 A Novel (Permitting Space Resource Extraction) Interpretation of Article II of the Outer Space Treaty

According to the US State Department,

> The Artemis Accords, which are grounded in the Outer Space Treaty of 1967, establish a common framework to guide responsible space exploration. The Accords' principles reflect the signatories' mutual dedication to the responsible and sustainable exploration and utilization of space.[59]

These intentions can hardly be questioned. However, a closer look reveals in Section 10 of the Accords ("Space resources") an alternative interpretation of Article II of the 1967 Outer Space Treaty, according to which the exploitation of space resources does not contradict the principle of non-appropriation. The whole section 10 reads as follows:

SECTION 10 – SPACE RESOURCES

1. The Signatories note that the utilization of space resources can benefit humankind by providing critical support for safe and sustainable operations.
2. The Signatories emphasize that the extraction and utilization of space resources, including any recovery from the surface or subsurface of the

---

[57] "The distinction between a transaction which is a definitive legal commitment between two states, and one which involves something less than that, is difficult to draw. The form or title of the instrument, for example a joint communiqué, is not decisive" – J. Crawford, Brownlie's Principles of Public International Law, Oxford University Press, 9th Edition, 2019, p. 353 et seq.

[58] See, for instance, that while Matusiak notes the political nature of the agreements, he ends by emphasising that "It may provide a legal ground for an activity of a few states in their common goal" – M. Matusiak, "Artemis Accords – A new era of space law or an initiative of the World's superpower?", Studia Juridica, XCVII, 2023, p. 182. In a broader sense, legal theory is somewhat embarrassed about characterizing the Artemis Accords. Neef, for example, notes that "the Artemis Accords (and any future agreements like this) provide a potential new path forward for creating the law in outer space, whereby dominant space faring nations seek to use their soft power to push forward interpretations of the law that benefit their own interests" (Neef, op. cit., p. 579). Or, that «they sit at the cross roads between global multilateral and unilateral…, but their status is further complicated by their political character" (Neef, *op. cit.*, p. 576).

[59] https://www.state.gov/first-meeting-of-artemis-accords-signatories/ (last accessed: 30.07.24).

Moon, Mars, comets, or asteroids, should be executed in a manner that complies with the Outer Space Treaty and in support of safe and sustainable space activities. The Signatories affirm that the extraction of space resources does not inherently constitute national appropriation under Article II of the Outer Space Treaty, and that contracts and other legal instruments relating to space resources should be consistent with that Treaty.

3. The Signatories commit to informing the Secretary-General of the United Nations as well as the public and the international scientific community of their space resource extraction activities in accordance with the Outer Space Treaty.

The meaning of paragraph 1 is not entirely clear. At first glance, it seems to refer to the use of resources *in situ,*[60] assuming that by 'operations' one means space operations. Even so, the concept of operations, as used here, remains unclear.

In this respect, the first sentence of the second paragraph refers to 'extraction and utilization of space resources', which should take place in a manner 'that complies with the Outer Space Treaty'—without, however, specifying how this will be ensured—'in support of safe and sustainable space activities'. These 'activities' could be the 'activities conducted in support of the Artemis Program' of the Preamble and thus this term would correspond to the 'safe and sustainable operations' of the first paragraph. It could then be assumed that the sentence in question also refers to *in situ* resource utilization (e.g. to serve facilities on the Moon or for the use of further space missions from its surface). However, this would be a restrictive interpretation, which it is doubtful that the drafters of the paragraph intended, given the broader context announced in the Preamble by 'RECOGNIZING the global benefits of space exploration *and commerce*'.

The second sentence of the second paragraph is the core of the whole section:

> The Signatories affirm that the extraction of space resources does not inherently constitute national appropriation under Article II of the Outer Space Treaty, and that contracts and other legal instruments relating to space resources should be consistent with that Treaty.

This is essentially an interpretation of Article II OST, according to which the extraction of space resources is not appropriation within the meaning of that article. Said interpretation basically reflects the position of the United States on the issue of space resources,[61] has the absolute character of a statement without a corresponding legal basis, and is in clear contradiction with the second half of the proposal, since nowhere does it show consistency with the Outer Space Treaty.[62]

---

[60] See, in this respect, Hobe, *op. cit.*, p. 7.

[61] "...this section is a clear example of the U.S. using the Accords to advance its interpretation of the Outer Space Treaty under international law" – W.A. Smith, "Using the Artemis Accords to Build Customary International Law: A Vision for a U.S.-Centric Good Governance Regime in Outer Space", Journal of Air Law and Commerce, vol. 86, no. 4, 2021, p. 683.

[62] "Ultimately, the Accords represent a compromise. They do not expressly state that space resource extraction is legal. Rather, they simply state a negative: that such activity would not in and of itself

The only reference that is made here is to the 'extraction' of space resources. This is even though the beginning of the paragraph reads 'the extraction *and utilisation* of space resources, including any recovery from the surface or subsurface of the Moon'. In the author's view, this is a deliberately incomplete approach, since in its complete form it would be in obvious contradiction with Article II, and possibly more difficult to accept. Said integrated form is meant to read 'extraction *and utilisation* of space resources', since no extraction—not only on celestial bodies, but also on Earth—makes sense in itself, but is always carried out with some purpose of exploitation/utilisation of the product of extraction.[63] Otherwise, an *ad litteram* interpretation of the term 'extraction of space resources' either would leave "the meaning ambiguous or obscure" or would "lead[s] to a result which is manifestly absurd or unreasonable", in the terms of Article 32 of the VCLT.[64] Always under this provision, these interpretative uncertainties can be overcome by taking into account the "circumstances of [its] conclusion" of the Artemis Accords, which are certainly determined by the broader position of the United States on the matter, as it has been expressed over many years, both at the legal/legislative and at the political level.[65] For instance, the emphasis on the use and acquisition of property rights over space resources is evident in the U.S. Commercial Space Launch Competitiveness Act,[66] which, states, *inter alia*, that:

> H.R.2262 - U.S. Commercial Space Launch Competitiveness Act.
> 114th Congress (2015-2016)
> TITLE IV--SPACE RESOURCE EXPLORATION AND UTILIZATION
> Space Resource Exploration and Utilization Act of 2015
> (Sec. 402) The bill directs the President, acting through appropriate federal agencies, to:

---

amount to national appropriation" - Jack W. Nelson, "The Artemis Accords and the Future of International Space Law", 24 ASIL INSIGHTS 1, 3 (2020), cited by Osada, op. cit., p. 420.

[63] Even an extremely small-scale mining operation that takes place for purely scientific reasons has a purpose for which it takes place!

[64] Article 32

Supplementary means of interpretation

Recourse may be had to supplementary means of interpretation, including the preparatory work of the treaty and the circumstances of its conclusion, in order to confirm the meaning resulting from the application of article 31, or to determine the meaning when the interpretation according to article 31:

(a) leaves the meaning ambiguous or obscure; or

(b) leads to a result which is manifestly absurd or unreasonable.

Hobe is in favor of a restrictive approach – Hobe, op. cit., p. 8.

[65] After all, as Hobe points out, the Artemis Accords "is a document formulated by one State to offer a common basis for cooperation between the contracting partners" – Hobe, op. cit., p. 4. See also the reasoning of F. Tronchetti & H. Liu, "Australia's Signing of the Artemis Accords: A Positive Development or a Controversial Choice?", AUSTL. J. INT'L AFF., Vol. 75, 2021, p. 246.

[66] "...from the perspectives of strategies and approaches, the Accords are in line with the Space Resource Exploration and Utilization Act of 2015 and the Presidential Decree No. 13914 in April 2020" – G. Wang, "NASA's Artemis Accords: the path to a united space law or a divided one?", The Space Review, Monday, August 24, 2020, https://www.thespacereview.com/article/4009/1/.

- facilitate *the commercial exploration for and commercial recovery of space resources* by U.S. citizens;…

> A U.S. citizen engaged in commercial recovery of an asteroid resource or a space resource *shall be entitled to any asteroid resource or space resource obtained, including to possess, own, transport, use, and sell it* according to applicable law, including U.S. international obligations (emphasis added).

It follows that Section 10 of the Artemis Accords essentially contains an interpretation of Article II OST, according to which the extraction and use of space resources does not constitute appropriation.

Finally, according to the third paragraph of the section, the States Parties to the Accords undertake to inform the United Nations Secretary-General "of their space resource extraction activities in accordance with the Outer Space Treaty". Article XI OST provides for an obligation to inform the Secretary-General, "of the nature, conduct, locations and results of... activities", "In order to promote international co-operation in the peaceful exploration and use of outer space". Obviously, Article XI refers to space activities which are in accordance with the OST and not of questionable international legal status.

5.2.2.1 Is This an Authentic Interpretation?

This interpretation cannot be regarded as an authentic interpretation of Article II of the Outer Space Treaty (OST). This would presuppose either a "subsequent agreement between the parties regarding the interpretation" of the OST or a "subsequent practice in the application of the [OST] which establishes the agreement of the parties regarding its interpretation", as set forth in Article 31 para. 3 (a) and (b) VCLT. In other words, it would necessitate an interpretative agreement (or a corresponding practice demonstrating an interpretative approach) of *all* States parties to the OST.[67] However, this is not the case, since, at least to date, (only) 54 States have signed the Artemis Accords.[68]

[67] "An authentic interpretation… emanates from the entity that created the interpreted norm" – K. Berner, "Authentic Interpretation in Public International Law", ZaöRV, Vol. 76, 2016, p. 846–847. See also Dörr & Schmalenbach, who suggest that both paragraphs 2 and 3 of Article 31 VCLT may lead to an authentic interpretation – O. Dörr & K. Schmalenbach (Eds.), Vienna Convention on the Law of Treaties - A Commentary, Springer, 2nd Edition, 2018, p. 590, 593–594, as well as Villiger, *op. cit.*, p. 429.

[68] See in this respect Deplano, op. cit., p. 807. As Kai-Uwe Schrogl mentions, "[United States] is aware that this cannot constitute an authoritative interpretation of [the space] treaties and it cannot constitute a new universal international space law either" - "We Must Not Overrate the Artemis Accords", in: SpaceWatch Global. https://spacewatch.global/2020/11/spacewatchgl-interviews-kai-uwe-schrogl-we-must-not-overratethe-artemis-accords/.

5.2.2.2 Is This an Acceptable Evolutive Interpretation?

It could also be argued that Section 10 constitutes an evolutive interpretation of Article II OST. Evolutive interpretation in international treaty law is intended to adapt the interpreted concept to contemporary reality. In other words, "an international instrument has to be interpreted and applied within the framework of the entire legal system prevailing at the time of the interpretation".[69] The interpretation of Article II OST contained herein is considered to be evolutive, given that, at the time of the adoption of the OST, the use and exploitation of space resources was not an issue under discussion, and that the prevailing interpretation of the scope of the non-appropriation principle of Article II includes private appropriation. The evolutive character becomes evident when one considers that the adoption of the Moon Agreement in 1979 (MOON) in the past constituted a kind of "subsequent agreement" or "subsequent practice" (depending on one's point of view), albeit implicit, of States on the interpretation of Article II OST in relation to the exploitation of space resources. The MOON was adopted by COPUOS by consensus[70] and was commended by the UN General Assembly through Resolution 34/68.[71] Obviously, this is not an authentic interpretation of the OST either, since, although it is in force, it has only been ratified by 17 (now) States, but it is (was) also of an evolutive nature, compared to the considerations of 1967.

Combining Articles 7 and 11 (2), (3) and (5), what the MOON provides on the issue of space resource exploitation can be summarized as follows:

The international regime for the exploitation of resources provided for in Article 11 para. 5 does not constitute an appropriation within the meaning of Article 11 paras. 2 and 3 (the prohibition of which clearly and explicitly includes that of private ventures), nor does it constitute a disruption of the existing balance of the (lunar) environment, as defined in Article 7.[72]

---

[69] See ICJ, Legal Consequences for States of the Continued Presence of South Africa in Namibia (South West Africa) notwithstanding Security Council Resolution 276 (1970), Advisory Opinion, I.C.J. Reports 1971, p. 16, at p. 31. In another case, the ICJ justified the evolutive interpretation by mentioning that when a "generic term" is used "denoting any matters comprised within the concept… the presumption necessarily arises that its meaning was intended to follow the evolution of the law and to correspond with the meaning attached to the expression by the law in force at any given time' – ICJ, Aegean Sea Continental Shelf, Judgment, I.C.J. Reports 1978, p. 3, at p. 32.

[70] Report of the Committee on the Peaceful Uses of Outer Space, COPUOS 22nd session, 18 June-3 July 1979, pars. 55–66.

[71] UN General Assembly, Resolution 34/68, Agreement Governing the Activities of States on the Moon and Other Celestial Bodies, 89th plenary meeting, 5 December 1979.

[72] See in this respect paragraph 65 of the COPUOS Report of 1979: "65. Following a suggestion for further clarification of article VII, the Committee agreed that article VII is not intended to result in prohibiting the exploitation of natural resources which may be found on celestial bodies other than the earth but, rather, that such exploitation will be carried out in such a manner as to minimize any disruption or adverse effects to the existing balance of the environment" - Report of the Committee on the Peaceful Uses of Outer Space, COPUOS 22nd session, 18 June-3 July 1979.

However, it is emphasized that, according to the MOON, only such an international regime constitutes an exception to the non-appropriation principle, and not any other relevant activity, against which the non-appropriation principle regains its full scope.

It should be noted, of course, that evolutive interpretation is not a panacea; as the International Court of Justice has stressed, it depends on the "presumption" that, when "a generic term is used", "…its meaning was intended to follow the evolution of the law and to correspond with the meaning attached to the expression by the law in force at any given time".[73] In the present case, this presumption must be demonstrated and the specific evolution of the law established, which is not readily apparent. Moreover, apart from an authentic interpretation, the interpretation of treaty provisions made by States parties to a treaty (not all of them) is binding only on the States making it. Its importance is therefore relative.

5.2.3 *Artemis Accords* vs. *the Moon Agreement: Conflicting Treaty Obligations*

An interesting issue that arises in relation to the Artemis Accords has to do with those countries that are simultaneously parties to both the Moon Agreement and the Artemis Accords. As of July 2024, this question concerns the following States: Armenia, Australia, Austria, Belgium, Chile, Mexico, the Netherlands, Uruguay and Peru. It also affected Saudi Arabia, but that country gave notice of its withdrawal from the Moon Agreement to the United Nations on 5 January 2023, and its withdrawal took effect on 5 January 2024.[74] Given the different approaches of these two instruments to the issue of space resources,[75] as discussed above, the question arises as to whether there is a conflict between the relevant obligations under these documents and, if so, how this conflict can be resolved.

As noted above, under the Moon Agreement, only the provision of Article XI para. 5 (with the characteristics provided for in Article XI para.7) does not constitute an appropriation, whereas, according to Section 10 of the Accords, the (any) extraction of space resources (by any entity, government or private person) does not constitute national appropriation under Article II of the Outer Space Treaty. There is an essential difference between the two approaches, since the Moon Agreement accepts the exploitation of resources only in the context of an international regime

[73] ICJ, Aegean Sea Continental Shelf, Judgment, I.C.J. Reports 1978, p. 3, at p. 32.

[74] https://treaties.un.org/doc/Publication/CN/2023/CN.4.2023-Eng.pdf (last accessed: 27.07.2024).

[75] See Mosteshar, who mentions that, among the principles contained in the Artemis Accords, there is "acceptance of the United States interpretation of Article II of the Outer Space Treaty as permitting appropriation of space resources and consequent rejection of the provisions of the Moon Agreement as part of general international law that would be binding on States not parties to the Moon Agreement" = S. Mosteshar, "Artemis: The Discordant Accords", Journal of Space Law, vol. 44, No 2, 2020, p. 596.

with specific characteristics, and therefore exceptionally in relation to the prohibition of appropriation in Article II of the OST, which is only bent before such a regime, while the Artemis Accords accept that *any* extraction (even unilateral, by a State or by a private entity) does not constitute appropriation. It is therefore clear that we are dealing with two substantially different interpretations of the concept of 'exploitation of space resources—non-appropriation', since one (the Moon Agreement) is strictly exceptional, while the other (the Artemis Accords) is perfectly permissible.[76,77]

In this case, the provision of Article 30 para. 4 VCLT (Application of successive treaties relating to the same subject matter) shall be applicable. Obviously, the concept of 'successive treaties' here does not refer to the entire body of the relevant instruments, but to Article 11 of the Moon Agreement in relation to Section 10 of the Artemis Accords, which constitute 'successive treaty provisions relating to the same subject matter'.[78] Pursuant to Article 30 VCLT,

> Article 30
> Application of successive treaties relating to the same subject matter
>
> 1. Subject to Article 103 of the Charter of the United Nations, the rights and obligations of States Parties to successive treaties relating to the same subject matter shall be determined in accordance with the following paragraphs.
>
> …
>
> 3. When all the parties to the earlier treaty are parties also to the later treaty but the earlier treaty is not terminated or suspended in operation under article 59, the earlier treaty applies only to the extent that its provisions are compatible with those of the later treaty.
> 4. When the parties to the later treaty do not include all the parties to the earlier one:
>
> (a) as between States Parties to both treaties the same rule applies as in paragraph 3;
> (b) as between a State party to both treaties and a State party to only one of the treaties, the treaty to which both States are parties governs their mutual rights and obligations.

It follows, according to the above, that "different speeds" of engagement are created: between States that are parties to both the MOON and the Artemis Accords, the interpretative approach of the Artemis Accords prevails and applies. Between States that are parties to both treaties and those that are parties only to the MOON,

---

[76] "Treaties are incompatible with each other if their obligations cannot be complied with simultaneously, i.e. if a State Party to both treaties cannot comply with one of them without breaching the other" – Dörr & Schmalenbach, *op. cit.*, p. 545.

[77] "On their face, the Accords seem to conflict directly with the Moon Agreement" – Smith, *op. cit.*, p. 683. See, *contra*, Ricky J. Lee, "Saudi Arabia's Withdrawal from the Moon Agreement", Australian and New Zealand Society of International Law, 25 July 2023, https://anzsilperspective.com/saudi-arabias-withdrawal-from-the-moon-agreement/.

[78] In other words, conflict clauses – see Villiger, *op. cit.*, p. 404.

the relevant interpretation of the latter will continue to apply. It is, however, obvious that the situation created by the different levels of commitment is not conducive to security in international relations on the issue of the exploitation of space resources, since, for States that are now parties to both instruments, it is logical that parallel participation should create a kind of ambiguity as to their position on this issue.[79,80] Therefore, from a legal point of view, it would be preferable for the States Parties to the MOON either not to become parties to the Artemis Accords or, if they do, to withdraw from the MOON at the same time—as Saudi Arabia has done.[81]

***

Regarding the issue of space resources, and beyond the theoretical debate on their binding or non-binding nature, the Artemis Accords essentially contain an interpretation of Article II OST that places space resource extraction absolutely beyond the principle of non-appropriation.[82] This approach contrasts with the restrictive interpretation of the 1979 Moon Agreement, which referred exclusively to an international regime for the exploitation of such resources.[83]

[79] "Para 4 lit b [VCLT]… does not really solve a conflict but instead inevitably leads to the breach of one of the treaties" - Dörr & Schmalenbach, op. cit., p. 552.

[80] As admitted by R.J. Lee, "A State cannot sign a later inconsistent, albeit non-binding, agreement and assert that it continues to observe the earlier treaty in good faith" – R.J. Lee, *op. cit.* For their part, and in relation to Australia's parallel participation in MOON and the Artemis Accords, Tronchetti & Liu note that "An additional problem for a country which has joined both the Moon Agreement and the Artemis Accords, is the fact that the former requires Parties to negotiate an international regime as soon as the exploitation of resources becomes feasible. Not only Australia has not entered into such negotiations with the other Parties to the Moon Agreement but, on the contrary, has joined an initiative of a country that has specifically rejected the Agreement itself" – *op. cit.*, p. 246–247.

[81] "Saudi Arabia… understands its continued support for the Moon Agreement is not congruent with its support of the Artemis Accords and therefore not in its state interest to remain a party" – M.J. Listner, "Four key points regarding Saudi Arabia's withdrawal from the Moon Agreement", The Space Review, Monday, December 11, 2023, https://www.thespacereview.com/article/4706/1/ (last accessed: 06.08.2024). Tronchetti & Liu are on the same line when they recommend "Australia to either reconsider its membership to the Moon Agreement or, at least, to clarify how the two instruments may co-exist[s] from a legal and political perspective" – *op. cit.*, p. 243.

[82] As Mosteshar comments, "it appears that through the Artemis Accords the United States is inviting likeminded States to join in freeing themselves of the appropriation prohibitions of the Outer Space Treaty" – Mosteshar, *op. cit.*, p. 602. The legal uncertainties of the Artemis Accords are also mentioned by Larsen, who has a generally positive view of the agreements – see P.B. Larsen, "Is There a Legal Path to Commercial Mining on the Moon?", University of Pittsburgh Law Review, vol. 83, no. 1, Fall 2021, p. 42 *et seq*. Smith agrees that Section 10 is "by far the most controversial, principle outlined in the Accords" – W.A. Smith, "Using the Artemis Accords to Build Customary International Law: A Vision for a U.S.-Centric Good Governance Regime in Outer Space", Journal of Air Law and Commerce, vol. 86, no. 4, 2021, p. 682.

[83] It is true that there is a difference in terminology between the two texts: the Moon Agreement speaks of exploitation, while the Artemis Agreement refers to extraction; however, it is difficult to imagine that the extraction of space resources will be done simply for the sake of doing it, with no ultimate purpose of exploitation, which obviously excludes in situ utilization.

What are the implications of this interpretation?

It is accepted in international law that international treaties may in some cases serve as precedents and contribute to the formation of customary law. This is particularly the case for treaties, whose purpose is not so much to create new rules as to clarify existing law.[84] The Artemis Accords are suitable for this model. Further, international law recognizes the equal legal value of custom and treaty as sources. Consequently, a custom may override or modify a treaty.[85] In accordance with applicable international law and the equality of its fundamental sources, namely treaty and custom, the proliferation of States signing and ratifying the Artemis Accords could, in the future, be the cause of the fact that such an interpretation of Article II of the 1967 Outer Space Treaty, authorizing the extraction (and exploitation) of space resources, will be firmly rooted in state practice and should therefore be legally accepted as having acquired customary force.[86]

The feasibility of such a process has been demonstrated by the International Court of Justice, in the *North Sea Continental Shelf* cases, which has shown the case of a "norm-creating provision [in a treaty] which has constituted the foundation of, or has generated a rule which, while only conventional… in its origin, has since passed into the general corpus of international law, and is now accepted as such by the *opinio juris*, so as to have become binding even for countries which have never, and do not, become parties to the [relevant] Convention. There is no doubt that this process is a perfectly possible one and does from time to time occur: it constitutes indeed one of the recognized methods by which new rules of customary international law may be formed…".[87]

In the context of this view, it is obvious that the decisive element here is numbers: The more countries that adhere to the Artemis Accords, the more the formation of a customary rule reflecting the specific interpretation of Article II OST—through the formation of an *opinio juris* favourable to it—is likely to emerge.[88]

This is, indeed, where the importance of the Artemis Accords lies, since for the rest they are based on and reproduce the fundamental principles of the 1967 Outer Space Treaty, except for the provision for the possibility of unilaterally establishing 'safety zones' in outer space (Section 11 par. 7), the international legitimacy of which is questionable.[89] Regarding the legality of the security zones described here,

[84] Charles Rousseau, Droit international public, Tome I – Introduction et sources, Sirey, Paris, 1970, p. 334.

[85] *Idem*, p. 342 *et seq.*

[86] Also see, in this respect, F.G. von der Dunk, "The Artemis Accords as a Tool of Cooperation", Proceedings of the International Institute of Space Law, 2021, p. 156.

[87] ICJ, *North Sea Continental Shelf*, Judgment, I.C.J. Reports 1969, p. 3, paras. 71–72.

[88] For the importance of the range of participation in the agreements, see M. Manoli, The Immaterial Territoriality of Outer Space: Reconstructing Territorylessness in Modern Space Governance (Montreal: McGill University, 2021), online: EScholarship https://escholarship.mcgill.ca/concern/theses/hd76s530m, p. 215.

[89] Something that R.J. Lee has noticed as well, although he seems to be generally supportive of these zones – Lee, *op. cit.* Also see M. Matusiak, *op. cit.*, p. 182.

the comments made in the context of the Hague Building Blocks also apply in this respect.

5.3 The International Lunar Research Station

The International Lunar Research Station (ILRS) is a project led by China and Russia.[90] Alongside the US-led Artemis Accords, the ILRS also seeks to implement international cooperation to return humans to the Moon. As its co-founders note, "the most efficient and productive investigation, exploration and use of the Moon can be achieved only in a broad international partnership with an attraction of other countries, international organizations and international partners… CNSA and ROSCOSMOS jointly invite all interested international partners to cooperate and contribute more for the peaceful exploration and use of the Moon in the interests of all humankind, adhering to the principles of equality, openness and integrity".[91]

As of 5 September 2024, 11 more countries (in addition to the co-founders, China and Russia) have joined the project (Azerbaijan, Belarus, Egypt, Kazakhstan, Nicaragua, Pakistan, Senegal, Serbia, South Africa, Thailand, Venezuela).[92] With a start date of 2021, the project is on track for completion in 2035 and will be in use from 2036.

ILRS is defined as "a complex experimental research facilities to be constructed with a possible attraction of partners on the surface and/or in the orbit of the Moon designed for multi-discipline and multi-purpose scientific research activities, including exploration and use of the Moon, moon-based observation, fundamental research experiments and technology verification, with the capability of long-term unmanned operation with the prospect of subsequent human presence".[93] The main purposes of the project include lunar topography, geomorphology and geological structure, lunar physics and internal structure, lunar chemistry (materials and geochronology), cis-lunar space environment, lunar-based astronomical observations, lunar-based Earth observations, and lunar-based biological and medical experiment. Regarding space resources, reference is (only) made to "lunar resources in-situ utilization". Thus, commercial exploitation of resources seems to be outside the scope of ILRS activities, which are mainly presented as having a scientific connotation, at least at the level of stated intentions.

According to the planned set of missions (2031–2035), Mission 2 includes, *inter alia*, lunar sample collection, while Mission 3 consists of the establishment of lunar

---

[90] …through the signing of a "Memorandum of Understanding between the Government of the People's Republic of China and the Government of the Russian Federation Regarding Cooperation for the Construction of the International Lunar Research Station", on 9 March 2021.

[91] https://www.cnsa.gov.cn/english/n6465652/n6465653/c6812150/content.html (last visit on 16 August 2024).

[92] https://swfound.org/lunar-space-cooperation-initiatives/ (last accessed: 18.04.25).

[93] *Idem*.

in-situ resource utilization technology verification facilities.[94] Thus, at least according to the current planning, ILRS is focused on the *in situ* use of lunar resources, and not on any form of commercial exploitation.

***

This Part has examined how different approaches to the search for a legal framework governing the exploration and use of space resources are reflected in international practice. The Hague Building Blocks constitute a *de lege ferenda* legal model, the implementation of which would require significant departures from the international space law currently in force, while at the same time not appearing to advance the prospect of an equitable sharing of such resources.

The Artemis Accords—proposed by the United States and gradually receiving positive responses from an increasing number of States—suggest an interpretation of Article II of the Outer Space Treaty under which the extraction of space resources does not constitute a breach of the principle of non-appropriation. The practical implications of this interpretation have been discussed above.

The International Lunar Research Station (ILRS), by contrast, is primarily a Moon return project, addressing the *in-situ* use of lunar resources only incidentally.

Taken together, these initiatives reflect the strong interest of States—and, more broadly, of space stakeholders—in the issue of space resources. However, despite their relevance, none of them amounts to a comprehensive solution. For various reasons, they do not appear to converge into a coherent legal model that ensures adequacy, completeness, and global acceptance. For this to happen, a process of reviving international law will be necessary.

[94] *Ibid.*

Chapter 6
Taking a Firm Step: Reviving International Law

It is clear from what has already been said, that an effective, *ad hoc* international regulation of the issue of space resources is needed. It is time for international law to fully address this issue. International law—and in particular international space law—needs to be updated in this regard, so that international relations surrounding the exploration and exploitation of space resources can enjoy the necessary security to make such activities feasible and effective. The new realities in the field of space activities, and, among them, the promising prospect of exploitation of space natural resources, the particular interest of the private sector in this area, the discussions on commercial exploitation, all these factors together create a new situation that cannot be adequately addressed by the international law of outer space of the 1960s and 1970s.

The legal mechanisms through which the desired revival of international law will be achieved will be determined by the will of the States. In this respect, the fifth—and final—part of this book attempts to explore the possible legal routes through which international law can take a firm step towards resolving the space resource issue.

6.1 Re-inventing or Re-positioning International Law

International law will either must be reinvented through the adoption of new, adapted rules or repositioned, through the emergence of new customary rules.

G. D. Kyriakopoulos, *Space Resource Utilization and Exploitation: The Suspended Step of International Law*, Space and Society,
https://doi.org/10.1007/978-3-031-96336-0_6

6.1.1 *Re-inventing International Law: Adopting New,* Ad Hoc *International Regulations*

Such a "new" law will either be made up of a framework of new binding rules or, if it fails to emerge as a common denominator of States' perceptions—and given the need for consensus in decision-making within the COPUOS—, it will again be limited to a set of principles or guidelines of a non-binding nature, i.e. following the pattern mainly followed by space law in the last decades of its evolution, after the 1979 Moon Agreement.

6.1.1.1 The "Hard Law" Approach: Adoption of an International Treaty

As already mentioned, the existing international conventional framework for outer space activities, in essence (but also in historical terms), illustrates the optimal balancing of the positions of the United States and the Soviet Union during the Cold War era. The interaction between the two superpowers of the time was, at first, reflected in a set of fundamental principles initially formed in the context of the Resolution 1962 (XVIII)/1963 of the General Assembly of the United Nations and finally incorporated in the Treaty on Principles Governing the Activities of States in the Exploration and Use of Outer Space, including the Moon and Other Celestial Bodies of 1967. These principles, which—as already presented in the course of this book—were destined to determine, decisively and over time, the identity of the current international law of outer space, include: the freedom of exploration and use of outer space (which constitutes a "province of all mankind") by all States without discrimination; the freedom of access to all areas of celestial bodies; the principle of non-appropriation of outer space; the use of the Moon and other celestial bodies exclusively for peaceful purposes; the rescue and return of astronauts in distress; the establishment of State responsibility with respect to national activities, whether such activities are carried on by governmental agencies or by non-governmental entities; State liability for damages caused by space objects; and the preservation of the Earth and the outer space environment.

These principles were further elaborated through the adoption of four more treaties, namely: The Agreement on the Rescue of Astronauts, the Return of Astronauts and the Return of Objects Launched into Outer Space of 1968; the Convention on International Liability for Damage Caused by Space Objects of 1972; the Convention on Registration of Objects Launched into Outer Space of 1975; and the Agreement Governing the Activities of States on the Moon and Other Celestial Bodies of 1979.

As described above, the adoption of the Moon Agreement in 1979 marked the end of the conventional cycle of space law development, at least so far. From that point on, as will be shown below, the international space community chose the path of soft law arrangements to address important issues in the context of space activities that required specific regulation in a way that the above-mentioned principles were not sufficient on their own.

So, the question now is: will the conventional cycle be reopened? The answer to this question is particularly crucial for the issue of space resources, since the definition of an *ad hoc* legal regime for this purpose by means of an international treaty—to the extent that States can agree to it, of course—would provide maximum legal certainty.

***

A new conventional framework would be the optimal solution to the issue of the effective regulation of issues related to the utilization and exploitation of space resources. Technically, this could take various forms:

- Either in the form of protocols to amend specific provisions of the existing treaties in force, such as the principle of non-appropriation or/and others;
- Or as a stand-alone, *ad hoc* international treaty dedicated to the issue in question;
- Or, on the other hand, as a partial set of provisions in the context of a comprehensive, global treaty on space activities, which, possibly along the lines of UNCLOS, will replace all existing space treaties and constitute the new *Magna Carta* of Space law.

In this direction, there have already been relevant suggestions and proposals from scholars: Tronchetti, for instance, has proposed a "Draft International Agreement on the exploitation of natural resources of the Moon and other celestial bodies".[1] Said model instrument "establishes a legal regime to regulate the exploitation of natural resources of the Moon and other celestial bodies", whose purpose is to "to create the conditions for the peaceful, safe and orderly development of the exploitation of natural resources of the Moon and other celestial bodies".[2]

Further, this model Agreement follows the pattern of the exploitation of international Seabed, which was provided in Part XI of the 1982 United Nations Convention on the Law of the Sea (UNCLOS), and subsequently amended by the so-called 1994 Agreement relating to the Implementation of Part XI:[3] Article II para. 3 provides that "The exploitation of natural resources of the Moon and other celestial bodies is managed by an International Space Authority as set out by the present Agreement. The Authority provides States as well as private operators with the authorization to exploit the natural resources of the Moon and other celestial bodies in accordance with the procedure set out in Article XII and in Section V of the present Agreement".

As regards the legal status of the Moon and the celestial bodies, Tronchetti's draft agreement essentially echoes the 1979 Moon Agreement, since it provides that

---

[1] Fabio Tronchetti, The Exploitation of Natural Resources of the Moon and Other Celestial Bodies - A Proposal for a Legal Regime, Martinus NIjhoff, Leiden – Boston, 2009, p. 293 *et seq.*

[2] Draft International Agreement, article II paras. 1, 2. For the international seabed regime, see *infra*, under V.1.2.2.

[3] Agreement relating to the Implementation of Part XI of the United Nations Convention on the Law of the Sea of 10 December 1982, adopted on 28/07/1994, entered into force on 28/07/1996, 1836 UNTS 3, 33 ILM 1309 (1994), UN Doc. A/RES/48/263.

"The Moon and other celestial bodies are not subject to national appropriation or to appropriation by international intergovernmental or nongovernmental organization, national organization or non-governmental entity or any natural person…",[4] whereas "The exploitation of natural resources of the Moon and other celestial bodies shall be carried out in the interest and for the benefit of all countries".[5] It is also noteworthy that there is an emphasis on safeguarding the rights of both developing States and those that contribute particularly to the exploitation of space resources:

> The interests and the needs of developing States as well as the efforts of those States which have contributed to the exploitation of the Moon or other celestial bodies, as well as their rights to make profits and to generate a reasonable return on investment shall be given special consideration when organizing the exploitation of natural resources of the Moon and other celestial bodies.[6]

Finally, Section IV: Institutional mechanisms, taking as a model the legal regime for the exploitation of the international seabed, refers to an "International Space Authority", which is an international organization whose main organs are the Assembly, the Council and a Technical and Legal Committee.

* * *

However, if the international space community ultimately opts for the conventional mechanism as a tool for the comprehensive and adequate regulation of space resource activities, other alternatives could also be on the table. An interesting approach would be, for example, the universal acceptance, through a multilateral protocol in the Outer Space Treaty, of an alternative interpretation of the principle of non-appropriation, along the lines of that contained in the Artemis Accords, or, again, the introduction, through a treaty text, of a distinction between planets and other celestial bodies (asteroids, comets, etc.) in respect of which the extraction, utilization and exploitation of mineral resources would be permissible.

Such solutions will *de facto* require other changes to existing space law, since the granting of mining and exploitation rights in any legal form will affect the free access of all states to celestial bodies under Article I Outer Space Treaty or raise issues of planetary protection under Article IX Outer Space Treaty. In fact, it seems that broader, more comprehensive treaty solutions will prove to be more legally safe.

6.1.1.2 The "Soft Law" Approach: Adoption of a Set of Guiding Principles

Soft law has been the privileged law-making mechanism in space law in recent years. In fact, after the Moon Agreement, soft law has evolved into an exclusive method of adopting international legal rules. In the context of this trend, important

---

4 Draft International Agreement, Article III.

5 Draft International Agreement, Article VI para.1.

6 Draft International Agreement, Article VI para. 2

principles or guidelines have seen the light of day, such as the Principles Governing the Use by States of Artificial Earth Satellites for International Direct Television Broadcasting,[7] the Principles relating to remote sensing of the Earth from outer space,[8] the Principles Relevant to the Use of Nuclear Power Sources in Outer Space,[9] the UNCOPUOS Space debris mitigation guidelines[10] or the UNCOPUOS Guidelines for the Long-term Sustainability of Outer Space activities.[11] In particular:

Over the years, and after the (last, chronologically speaking) Moon Agreement of 1979, there has been a growing reluctance on the part of the international community (especially on the part of States with a strong space presence, such as the United States) to proceed with the adoption of new conventional texts to further regulate international space relations. It is clear that the recent developments in space exploration and utilization have created a challenging environment for the space community, particularly in terms of the increasing private sector activity and the growing commercialization of space endeavors. These changes have brought about a legal and regulatory dilemma for the space community with regards to the adoption of new binding provisions, particularly given the global common character of international space law in force. It was therefore to be expected that States would resort to more flexible approaches in order to advance regulation, by adopting non-binding ("soft law") provisions. These texts are usually referred to as 'principles' or 'guidelines'. They may also take the form of 'codes of conduct' or 'best practices'. The most important non-binding texts in the context of space activities are those already highlighted.

Although the above rules are not binding on States, consistent, continuous and uniform adherence to them can, as has already been argued for the General Assembly Resolutions, "signpost" the direction that the (future, conventional or customary) rule of law will take. Within this context, many scholars argue that the COPUOS Guidelines on Space Debris Mitigation have already attained the status of customary international law, given their uniform application by the major national space agencies, as well as by the European Space Agency (ESA).

However, it should be emphasized that the increasing intervention of the private sector in the space domain cannot be regulated in the long term by non-binding texts. Sooner or later, the international community will have to revisit the question of the further adoption of new contractual (binding) provisions, either by amending the existing space treaties or by negotiating new international agreements. Another solution that has been discussed from time to time is the adoption of a 'comprehensive' treaty on outer space, following the (successful) example of the 1982 United Nations Convention on the Law of the Sea.

---

[7] A/RES/37/92 of 10 December 1982.

[8] A/RES/41/65 of 3 December 1986.

[9] A/RES/47/68 of 14 December 1992.

[10] A/RES/62/217 of 22 December 2007.

[11] A/RES/74/82 of 26 December of 2019.

This is because, given the existence of a framework of principles that form the backbone of space law, it is not easy to make the necessary legal transgressions for an effective and appropriate regulation of space resource activities through "soft law" approaches. The non-binding principles or guidelines already adopted by the international space community have made it possible and have had a significant positive impact on space affairs, as they do not contradict the essence of "hard" space law. For example, it is difficult to imagine how the mechanism of adopting principles of a non-binding character could substantially modify or even reverse the universally accepted principle of non-appropriation.

Of course, the mandate of the Working Group on Legal Aspects of Space Resource Activities includes, *inter alia*, the adoption of non-binding principles:

> A. Mandate
>
> 8. The working group shall:
>
> … (b) Develop a recommended set of initial principles and, if appropriate, practical measures for such activities to ensure that they are carried out in a safe, sustainable and peaceful manner *and in accordance with international law* for consideration by the Committee and possible subsequent consideration by the General Assembly in a dedicated resolution;… (emphasis added)[12]

It is obvious that the principles discussed (and sought) here will operate *intra legem*, in other words, they will operate within the established principles of space law, which they will in a sense reproduce. Such a set of principles will undoubtedly be useful, as it will constitute a step towards "legal awareness" on the part of the international space community on the issue of the utilization and exploitation of space resources. However, it is highly doubtful whether such principles as the ones discussed here can lead to substantive rules on the matter, rules which at present may even appear to be *contra legem* in some respects.

6.1.2 Re-inventing International Law #2: In Search of Patterns for Inspiration

The suitability of potential models must be assessed in relation to the specific characteristics of space law. Examples illustrating both suitable and unsuitable approaches are examined below.

[12] Committee on the Peaceful Uses of Outer Space, Sixty-fourth session, Vienna, 25 August–3 September 2021, Proposal on the mandate, terms of reference and method of work of the working group established under the Legal Subcommittee agenda item entitled "General exchange of views on potential legal models for activities in the exploration, exploitation and utilization of space resources", Working paper submitted by the Chair and Vice-Chair of the working group, A/AC.105/L.326, 25 June 2021.

6.1.2.1 Models Not Considered to be Suitable: Establishing an "Exclusive Economic Zone" in Outer Space

The theoretical—at least for the moment—debate on the creation of an "exclusive space economic zone" stems from the establishment of the Exclusive Economic Zone (EEZ) in the context of the Law of the Sea, under the 1982 United Nations Convention on the Law of the Sea.[13]

6.1.2.1.1 The Legal Elements of the EEZ at Sea

The justification for the introduction of the EEZ in the context of the Law of the Sea has its roots in the practice of Latin American States in the 1970s to declare zones of sovereignty 200 nautical miles from their coasts in order to "conserve, develop and exploit the natural resources of the maritime zones adjacent to their coasts, of their seabeds and the subsoil thereof;".[14]

Articles 55–75 of Part V of the Law of the UNCLOS outline the characteristics of the Exclusive Economic Zone (EEZ). Accordingly, the EEZ is as a maritime zone which needs to be declared by the coastal State, and which extends up to "200 nautical miles from the baselines from which the breadth of the territorial sea is measured".[15]

In the context of the EEZ, the coastal State enjoys sovereign rights to manage, explore, exploit, and conserve the living and non-living resources located within the zone. In any case, the rights, whether sovereign or jurisdictional, and the related duties of coastal States cannot be intended to be absolute, since due regard to the rights and duties of other States is expressly provided for by Articles 55, 56(2), and 58 of the UNCLOS. Article 58(1) specifically envisages the freedom of navigation and overflight, the laying of submarine cables and pipelines, and 'other internationally lawful uses of the sea' connected with the above-mentioned freedoms.[16]

To date, over 130 States have claimed an EEZ. In principle, a coastal State shall, through agreements, enable other States to have access to the surplus of its allowable catch; in doing so, it shall give special consideration to landlocked, geographically disadvantaged, and least developed States.[17] Additionally, regarding enforcement, capability of other States is limited because the coastal State may take decisions determining the total allowable catch (TAC), the extent of its harvesting

[13] Convention on the Law of the Sea, adopted on 10 December 1982, entered into force on 16 November 1994. 1833/U.N.T.S./397 (hereinafter UNCLOS).

[14] See the Declaration of Montevideo on the Law of the Sea, signed at Montevideo, Uruguay, May 8, 1970, International Legal Materials, Volume 9, Issue 5, September 1970, pp. 1081–1083. See also Y. Tanaka, The International Law of the Sea, 3rd ed., Cambridge UP, 2019, p. 149.

[15] UNCLOS, Art. 57.

[16] Tanaka, *op. cit.*, p. 151 *et seq.*

[17] UNCLOS, Arts. 62, 69–70.

ability and the allocation of surpluses fall outside of the compulsory dispute settlement system set out in Part XV of the Convention.[18]

Under Article 56 of the UNCLOS, coastal States are also conferred three kinds of jurisdictional rights concerning (a) the establishment and use of artificial islands, installations, and structures; (b) marine scientific research; and (c) the protection and preservation of the marine environment. The first right is particularly relevant for the present analysis as it entails the coastal State's exclusive jurisdiction to construct and operate artificial islands, installations, and structures for economic purposes.[19] Special notice must be also given when constructing an artificial island or installations and the coastal State may, where necessary, establish safety zones around them.[20] Additionally, in relation to such islands and installations, coastal States have exclusive jurisdiction to enact custom, fiscal, health, safety, and immigration laws and regulations.[21]

According to Article 58 of the UNCLOS, other States enjoy the freedom of navigation and overflight, two rights typical of the high seas.[22] However, these rights are less extensive than those that can be exercised on the high seas. As mentioned above, within the EEZ other States shall pay due regard to the rights of the coastal State and shall comply with the laws and regulations that such a State has adopted pursuant to the UNCLOS.[23] Furthermore, a coastal State can take measures to ensure compliance by foreign vessels of such laws.

6.1.2.1.2 The Discussion on an "Exclusive Utilization Space" in Outer Space: Why Such a Zone Would Be of Minimal Importance for the Exploitation of Space Resources

The idea of the establishment of an "Exclusive Utilization Space" modelled on the EEZ comes from a paper by Hao Liu & Fabio Tronchetti.[24] These authors essentially propose the creation of such a zone within the so-called Near Space, which occupies the space above 18 km—the height up to which most aviation activities take place—and below 100 km, i.e. the height at which, according to Theodore von Kármán's theories, outer space begins. According to the analysis, the so-called "Exclusive Utilization Space" (EUS) should be distinguished from national airspace and outer space and should be established by the "underlying State". Said State should retain priority rights in using the UAS; further, it could enforce safety

[18] See A. E. Boyle, "Dispute Settlement and the Law of the Sea Convention: Problems of Fragmentation an Jurisdiction", ICLQ, vol. 46, 1997.

[19] UNCLOS, Art. 60.

[20] UNCLOS, Arts. 60 (3)–(4).

[21] UNCLOS, Art. 60 (2).

[22] UNCLOS, Arts. 58 (1), 87.

[23] UNCLOS, Art. 58 (3).

[24] H. Liu & F. Tronchetti, "The Exclusive Utiization Space: A New Approach to the Management and Utilization of the Near Space", U. Pa. J. Int'l L., Vol. 40:3, 2019, p. 537–574.

and security measures. UAS would mainly serve the deployment of high-altitude platforms which may also be placed by third States, subject to prior notification to and approval by the State concerned.[25]

Is there a way that the Exclusive Space Zone can provide solutions to the issue of space resources?

Firstly, it should be noted that this concept has inherent weaknesses, even when considered in light of the stated objectives of its initiators.

The rationale behind the acceptance and international legal regulation of the EEZ was a compromise between coastal States that claimed extensive territorial sea zones and maritime powers that consistently defend freedom of navigation. The essence of this compromise can be summarized as follows: the coastal State is granted sovereign rights (and not sovereignty) of a functional (economic) nature, while freedom of navigation is maintained with regard to maritime transit. It is not possible to apply this concept to outer space, however, due to the different nature of state rights in airspace (in comparison to maritime space)[26] and outer space. In international airspace law, there is no recognition of "intermediate" zones of sovereign rights. This is because airspace is simply divided into national and international. National airspace is that in which the underlying State has full and exclusive sovereignty (airspace over State territory or territorial waters). International airspace allows for freedom of overflight and the applicable rules are international.[27] Furthermore, the concept of State sovereignty and the notion of 'sovereign rights' are not applicable in outer space due to the freedom of use and exploration of space and the principle of non-appropriation.[28]

It is notable that, in contrast to the situation at sea, proximity has not been a factor in the development of legal rights in international airspace, whereas, in outer space, there is no proximity to consider. Therefore, the rationale for establishing an EEZ cannot be directly applied to outer space, including the concept of "Near Space". This is because "Near Space" is not a recognized area within the context of international space relations. It is not currently acknowledged as a distinct zone, separate from airspace or outer space.

In any case, the concept of Near Space is still evolving and cannot be defined solely by the height, which is the ceiling for aircraft flights. Furthermore, it should be noted that this height is not a fixed limit, as there are currently flying machines capable of reaching much higher altitudes. By way of illustration, the U-2 aircraft has exceeded 21.3 km (70,000 ft),[29] while supersonic aircraft can reach 30 km.[30] It

---

[25] Liu & Tronchetti, *op. cit.*, p. 564 *et seq.*

[26] See G. Kyriakopoulos,"Statut des espaces maritimes et régime juridique de la navigation aérienne, RGDIP, 2001/2, p. 307–329.

[27] Arts. 1 and 12 of the 1944 Chicago Convention on International Civil Aviation, Art. 87 UNCLOS.

[28] Arts. 1 para. 2, 2 OST.

[29] https://www.af.mil/About-Us/Fact-Sheets/Display/Article/104560/u-2stu-2s/ (last accessed: 28.10.2024).

[30] See "How high do planes fly?", https://www.ferrovial.com/en/resources/how-high-do-planes-fly/ (last accessed: 28.10.2024).

is also important to note that the legal definition and delimitation of outer space (from airspace) has consistently been an unresolved issue.

If, as this analysis suggests, the establishment of an "Exclusive Utilization Space", based on the maritime Exclusive Economic Zone, is not a viable option, even for activities in what some may consider Near Space, it is clear that this approach is unsuitable for the exploration and exploitation of space resources. This is because the relevant reference is to the mineral resources of celestial bodies, which are spatially outside the specific reference area.

6.1.2.2 Legal Models to be Considered: Towards a Space Analogue of the International Seabed Authority?

6.1.2.2.1 The International Seabed Regime

One international legal regime that may provide an insightful model for the regulation of space resources is that of the exploitation of international seabed resources. The conditions on the international seabed can be considered comparable to those of outer space, given that human access is, in both cases (and to the respective resources), extremely difficult due to objective circumstances. Furthermore, both areas lie beyond national jurisdiction, while the legal regime of the so-called "Area" concerns only mineral resources, which also needs to be regulated in the context of outer space.

Historically, the international seabed regime and the consideration of the seabed as the "common heritage of mankind" have acted as a deterrent to the prospect of developed, technologically advanced States being allowed to monopolise the exploitation of the seabed beyond the limits of the continental shelf. This would have been possible if the international seabed had been considered *res nullius*—and therefore subject to occupation—or if it had been considered a *res communis*, which would have led the exploitation of its mineral resources to be dealt with in the context of the freedoms of the high seas—, or if the progressive extension of the limits of the continental shelf, in accordance with the criterion of exploitability introduced by the 1958 Geneva Convention on the Continental Shelf, would have led to the exploitation of the international seabed becoming a matter for the coastal States.[31]

The UNGA Resolution 2749/1970, "Declaration of Principles Governing the Seabed and Ocean Floor, and the Subsoil Thereof, beyond the Limits of National Jurisdiction" was pivotal in the introduction of the concept of the "common heritage of mankind" for the international seabed. Said resolution solemnly proclaimed, *inter alia*, the 'common character' of the area in question, both in its positive and negative dimensions. According to the resolution, said area shall not be subject to appropriation by any means by States or persons, natural or juridical, and no State shall claim or exercise sovereignty or sovereign rights over any part thereof.

[31] Y. Tanaka, The International Law of the Sea, 3rd ed., Cambridge UP, 2019, p. 217–218.

Furthermore, the "Area" is to be used exclusively for peaceful purposes and in accordance with the international regime to be established.[32]

Based on the above, Declaration 2749/1970, has a selective affinity with the pertinent provisions of the Moon Agreement concerning the international regime for the exploitation of the natural resources of the Moon.

Following the above declaration, the legal regime for the exploitation of the international seabed is set out in Part XI of the 1982 United Nations Convention on the Law of the Sea (UNCLOS).

Part XI of the UNCLOS and the Annexes attached to it were adopted following lengthy negotiations, during which conflicting interests among various groups of States[33] were taken into account. This resulted in the creation of a particularly complex legal regime for the extraction of minerals from the international seabed, which, according to Part XI, constituted the "Area".

The fundamental legal implications of the designation of the Area and its resources as "common heritage of mankind" are set forth in Articles 136, 137, and 140 of the UNCLOS.

According to Article 136, "The Area and its resources are the common heritage of mankind"; Article 137 para. 2 stipulates that "All rights in the resources of the Area are vested in mankind as a whole, on whose behalf the Authority shall act..."—whereas Article 140 provides that "Activities in the Area shall …be carried out for the benefit of mankind as a whole, irrespective of the geographical location of States, whether coastal or land-locked, and taking into particular consideration the interests and …" and, further, that "The Authority shall provide for the equitable sharing of financial and other economic benefits derived from activities in the Area…".

The 1994 Agreement. Although the system of exploitation of the international seabed resources, as set forth in Part XI, was original and innovative, it encountered objections from industrialized States. The United States, for instance, did not ratify the UNCLOS due to concerns regarding Part XI. Over time, however, the US came to acknowledge that the Convention, in its entirety, constitutes customary international law, except for that Part.[34] Moreover, in the early 1980s, the United States and other countries (Federal Republic of Germany, France, Japan, USSR, Italy) began

---

[32] In this respect, see also the UNGA Resolutions: A/RES/2340(XXII) of 18 December 1967, A/RES/2467(XXIII) of 21 December 1968 and A/RES/2574 (XXIV) of 15 December 1969.

[33] See, for instance, R.R. Churchill & A.V. Lowe, The Law of the Sea, 3rd Edition, 1999, Manchester University Press, p. 228. Also, M. Bennouna, "Les droits d'exploitation des ressources minérales des océans", *in:* D. Bardonet & M. Virally (dir.), Le nouveau droit international de la mer, Pedone, Paris, p. 127, O. Bresard, "Les aspects financiers de l'exploitation des ressources de fonds marins internationaux", *in:* SFDI, Perspectives du droit de la mer à l'issue de la 3e Conférence des Nations unies, Colloque de Rouen, Pedone, 1984, p. 313, R.-J. Dupuy, L'océan partagé, Pedone, 1979, in particular Part IV (Les appropriations collectives: Le 'patrimoine commun de l'humanité'), p. 135–199. Also, in the same spirit, J.-P. Lévy, La Conférence des Nations Unies sur le droit de la mer (Histoire d'une négociation singulière), Pedone, 1983, 159 p.

[34] Tanaka, *op. cit.*, p. 227.

to implement national legislation governing the exploitation of the international seabed.[35]

To address the objections, primarily from the industrialized States, to Part XI, the UN Secretary-General initiated informal consultations,[36] which resulted in the adoption, on 28 July 1994, of the *Agreement relating to the implementation of Part XI of the United Nations Convention on the Law of the Sea of 10 December 1982* (hereinafter "the Agreement" or "the 1994 Agreement").[37]

The Agreement is essentially an amendment to Part XI of the UNCLOS. It seeks to approach the management of the ocean seabed on "market terms" in order to facilitate wider participation by States in its management.[38] The Agreement entered into force on 28 July 1996. In the event of a conflict between the 1994 Agreement and Part XI of UNCLOS, the former shall prevail. Furthermore, the 1994 Agreement and Part XI of UNCLOS must be applied and interpreted as a single instrument.[39]

The 1994 Agreement introduced a series of significant amendments to Part XI, the most significant of which are outlined below.

6.1.2.2.2 The Area, Its Resources and the Activities Carried Out Therein

The main features of this regime are that it considers the designated zone (the "Area") and the resources within it to be the "common heritage of mankind", that it governs specific activities within it, that its management is entrusted to an international intergovernmental organization (the "Authority"), and that it is clearly guided by the need to protect the marine environment.

Part XI of the UNCLOS provides regulation for "the seabed and ocean floor and subsoil thereof, beyond the limits of national jurisdiction", which constitute the "Area".[40] The resources of the Area that are the subject of the relevant regulation are defined as "all solid, liquid or gaseous mineral resources *in situ* in the Area at or beneath the seabed, including polymetallic nodules". Furthermore, "resources, when recovered from the Area, are referred to as 'minerals'".[41]

---

[35] Also, a Provisional Understanding Regarding Deep Seabed Matters was adopted in 1984 between Belgium, France, Federal Republic of Germany, Italy, Japan, the Netherlands, the United Kingdom and the United States – see Tanaka, *op. cit.*, p. 227. See also F. Xu & J. Su, "Towards a legal regime of benefits sharing for space mining: With some experience from the Area", Resources Policy, Volume 76, 2022, article id. 102627, https://doi.org/10.1016/j.resourpol.2022.102627.

[36] See, in this respect, S. Rosenne, "The perplexities of modern international law: general course on public international law", R.C.A.D.I., Vol. 291, 2001, p. 303, as well as http://www.un.org/Depts/los/convention_agreements/convention_overview_part_xi.htm.

[37] Adopted on 28 July 1994 (New York, A/RES/48/263), entered into force on 28 July 1996, 1836/NTS/3. See http://www.un.org/Depts/los/convention_agreements/convention_overview_part_xi.htm.

[38] See T. Scovazzi, "The evolution of international law of the sea: new issues, new challenges", *R.C.A.D.I.*, Vol. 286, 2000, p. 125.

[39] Art. 2 para. 1 of the 1994 Agreement.

[40] UNCLOS, Art. 1.

[41] UNCLOS, Art. 133.

The "activities in the Area" are those encompassing "all activities of exploration for, and exploitation of, the resources of the Area".[42] Activities in the Area shall be conducted in such a way as to ensure the "effective protection for the marine environment from harmful effects which may arise from such activities"[43]—a requirement that sets clear limits, of environmental character, on how the resources of the international seabed should be managed. As already mentioned, the Area and its resources are considered "the common heritage of mankind".[44]

The foregoing activities include drilling, dredging, excavation, disposal of waste, construction and operation or maintenance of installations, pipelines and other devices related to such activities.[45] Furthermore, according to the International Tribunal for the Law of the Sea (ITLOS), "shipboard processing immediately above a mine site of minerals derived from that mine site"[46] also forms part of the activities in question. However, these activities, as outlined in Article 153 UNCLOS, must be distinguished from the transporting, processing and marketing of minerals recovered from the Area, operations described in Annex IV to the Convention, which are nevertheless similarly carried out by the so-called Enterprise.[47] Activities in the Area may not harm the marine environment.[48]

"Exploration", "exploitation", "prospecting". The terms "exploration" and "exploitation" are not further defined in the UNCLOS. Such definitions may nevertheless be found in regulations adopted by the Authority, such as:

The Regulations on Prospecting and Exploration for Polymetallic Nodules in the Area;[49]

The Regulations on prospecting and exploration for polymetallic sulphides in the Area;[50]

The Regulations on Prospecting and Exploration for Cobalt-rich Ferromanganese Crusts in the Area.[51,52]

---

[42] UNCLOS, Art. 1 para. 3.

[43] UNCLOS, Art. 145.

[44] UNCLOS, Art. 136. The CHM concept was introduced into the international debate on the law of the sea in 1967, by Maltese Ambassador Arvid Pardo – see Tanaka, *op. cit.*, p. 218.

[45] UNCLOS, Art. 145.

[46] …as indicated in Annex III Art. 17 para. 2(f) of the UNCLOS. See ITLOS, … para. 88.

[47] See ITLOS, Responsibilities and obligations of States with respect to activities in the Area, Advisory Opinion, 1 February 2011, ITLOS Reports 2011, p. 10, at para. 84 (hereinafter "ITLOS Advisory Opinion").

[48] UNCLOS, Art. 145.

[49] Originally adopted on 13 July 2000, amended in 2013 (ISBA/19/C/17).

[50] Adopted on 7 May 2010 (ISBA/16/A/12/Rev.1).

[51] Adopted on 27 July 2012 (ISBA/18/A/11).

[52] As noted by the ITLOS, "The Regulations are instruments subordinate to the Convention, which, if not in conformity with it, should be interpreted so as to ensure consistency with its provisions. They may, nevertheless, be used to clarify and supplement certain aspects of the relevant provisions of the Convention" – ITLOS Advisory Opinion, para. 93.

In the Polymetallic Nodules Regulation, "exploration" is defined as:

> ...the searching for deposits of polymetallic nodules in the Area with exclusive rights, the analysis of such deposits, the testing of collecting systems and equipment, processing facilities and transportation systems, and the carrying out of studies of the environmental, technical, economic, commercial and other appropriate factors that must be taken into account in exploitation;

whereas "exploitation" means:

> the recovery for commercial purposes of polymetallic nodules in the Area and the extraction of minerals therefrom, including the construction and operation of mining, processing and transportation systems for the production and marketing of metals.

The same definitions are found in the other two Regulations.

As observed by the ITLOS, the terms "exploration" and "exploitation", as defined in the Regulations, have a broader scope than the term "activities in the Area", as outlined in the UNCLOS. This is because the Regulations further include processing and transportation of minerals in the definitions of "exploration" and "exploitation".[53]

The concept of "prospecting", which in the context of UNCLOS is distinguished from exploration and exploitation and constitutes a preliminary phase of exploration, is referenced as a related topic. The Polymetallic Nodules Regulation defines prospecting as follows:

> 'prospecting' means the search for deposits of polymetallic nodules in the Area, including estimation of the composition, size and distribution of deposits of polymetallic nodules and their economic values, without any exclusive rights;.

Given these pre-existing definitions, it is reasonable to assume that the definitions of space resources—particularly in the context of discussions within the COPUOS Legal Subcommittee—should build upon them, while duly taking into account the relevant analogies and differing circumstances.

6.1.2.2.3 The Role of the International Seabed Authority and of its Organs

In accordance with Part XI of the UNCLOS, the entity responsible for the management of the Area is the International Seabed Authority (the "Authority"). The Authority is an international intergovernmental organization whose role is to ensure the organization and the control of the activities in the Area, "with a view to administering the resources of the Area".[54] Authority's membership extends to all States Parties to the Convention.[55] Said international body was established on 16 November 1994, when the UNCLOS entered into force. It has the powers and responsibilities conferred on it by the Convention and may also have implicit powers to the extent necessary for the exercise of its functions in relation to activities in the Area.[56]

[53] ITLOS Advisory Opinion, para. 92.

[54] UNCLOS, Art. 157 para. 1.

[55] UNCLOS, Art. 156 para. 2.

[56] UNCLOS, Art. 157 para. 2.

In particular, the Authority, through its Assembly, has the power to adopt and approve rules, regulations and procedures on the equitable sharing of financial and other economic benefits derived from activities in the Area.[57] Adoption and approval by the Assembly is based on the recommendation of the Authority's Council.[58] Further, the Authority shall consider and adopt regulations and procedures in relation to prospecting, exploration and exploitation in the Area,[59] such as the aforementioned regulations, which form the so-called "Mining Code".[60] These regulations are binding on the members of the Authority.[61]

In accordance with the UNCLOS, the Authority was empowered to borrow funds.[62] This capacity for borrowing funds is effectively revoked by the 1994 Agreement (Annex, Section 1, para. 14), which stipulates that the administrative expenses of the Authority shall be met through the budget of the United Nations and, thereafter, by assessed contributions of its members.

The Assembly. The Assembly consists of all members of the Authority. Under the UNCLOS, the Assembly was regarded as the supreme organ of the Authority, vested with the power to establish general policies "in conformity with the Convention".[63] However, the 1994 Agreement introduced a significant amendment to this, stipulating that the general policies of the Authority shall thereafter be established by the Assembly in collaboration with the Council (Annex, Section 3, para. 1). Regarding decision making, the Assembly was empowered, under the UNCLOS, to make decisions on matters of procedure by a majority of the members present and voting. However, decisions on questions of substance required a two-thirds majority.[64] Following the 1994 Agreement, the general rule now is that decisions must be taken by consensus.[65]

The Enterprise. In accordance with Part XI of the UNCLOS, the Enterprise was initially the organ of the Authority responsible for the implementation of activities within the Area. In addition to undertaking direct activities, its objective was the

[57] UNCLOS, Art. 160 para. 2 (f) (i).

[58] UNCLOS, Art. 162 (2) (o) (i). According to the 1994 Agreement, rules, regulations and procedures on the equitable sharing of financial and other economic benefits derived from activities in the Area and the relevant decisions shall take into account recommendations of the Finance Committee [1994 Agreement, Section 9, para. 7 (f)].

In July 2023, the Finance Committee considered a draft proposal for the establishment of a "Common Heritage Fund" (ISBA/28/FC/4), in context with the implementation of the 2030 Agenda for Sustainable Development and the achievement of the Sustainable Development Goals – see https://www.isa.org.jm/equitable-sharing-of-benefits/ (last accessed: 31.1.2025).

[59] UNLCOS, Art. 160 para. 2 (f) (ii).

[60] https://www.isa.org.jm/the-mining-code/ (last accessed: 31.1.2025).

[61] Tanaka, op. cit., p. 222.

[62] UNCLOS, Art. 174 para. 1.

[63] UNCLOS, Art. 160 para. 1.

[64] UNCLOS, Art. 159 para. 7.

[65] However, if all efforts to reach a decision by consensus have been exhausted, the majority rule set forth in Article 159 para. 7 is reinstated. The Agreement has introduced corresponding amendments to the Council's decision-making procedure, which is now based on the principle of consensus.

transporting, processing and marketing of minerals recovered from the Area.[66] The Enterprise "enjoyed autonomy in the conduct of its operations" under the UNCLOS.[67] However, this autonomy was effectively repealed by the 1994 Agreement. Section 2 paras. 2 and 3 of the Annex to the Agreement stipulate that the Enterprise shall henceforth conduct its deep seabed mining operations through joint ventures, in accordance with "sound commercial principles". Furthermore, the Agreement states that "States Parties shall be under no obligation to finance any of the operations in any mine site of the Enterprise or under its joint-venture arrangements". This clearly downgrades the role and importance of the Enterprise, which was originally conceived—in the context of UNCLOS—as a "mechanism for collective exploitation" and makes the success of its work uncertain.

Entities operating in the Area. Activities in the Area shall be carried out either by the Enterprise or, in association with the Authority, by States Parties, or state enterprises or natural or juridical persons which possess the nationality of States Parties or are effectively controlled by them or their nationals, according to Article 153 para. 2 of the UNCLOS. This is the so-called "parallel system".

Annex III of the 1982 Convention established a comprehensive regulatory framework governing the financial terms of contracts for the exploration and exploitation of the Area. The system provided for the payment of considerable fees by contractors.[68] Nevertheless, the 1994 Agreement introduced a fairer and simpler system, more compatible with market laws and competition rules, through the introduction of a set of principles which now regulate the financial terms of contracts in a simpler way.[69]

Under the terms of the 1994 Agreement, contractors now also have the right to be involved in the so-called reserved areas. These areas are essentially designated by developed States at the stage of the application for exploitation rights and relate to half of the area nominated for exploitation, which must be "of equal estimated commercial value".[70] UNCLOS reserves areas for exploitation by the Enterprise or jointly with an interested developing State. The 1994 Agreement substantially revised this concept in favour of the applicants by providing that designation of a reserved area for the Authority "shall take place in connection with approval of an application for a plan of work for exploration or approval of an application for a plan of work for exploration and exploitation".[71] In addition, a contractor which has contributed a reserved area has now a right of first refusal to enter into a joint-

[66] UNCLOS, Art. 170 para. 1.

[67] See UNCLOS, Annex IV, Art. 2 para. 2: "2. Subject to paragraph 1, the Enterprise shall enjoy autonomy in the conduct of its operations".

[68] In accordance with Article 13 paras. 2 and 3 of Annex III, a fee of $US 500,000 was to be paid for each application, while the contractor was required to pay an annual fixed fee of $US 1 million from the date of entry into force of the contract.

[69] See Section 8 (para. 2) of the Annex to the 1994 Agreement.

[70] UNCLOS, Annex III – Basic Conditions of Prospecting, Exploration and Exploitation, Arts. 8 and 9.

[71] 1994 Agreement, Annex, Section 1, para. 10.

venture arrangement with the Enterprise for exploration and exploitation of that area. If the Enterprise has been idle for 15 years, said contractor "shall be entitled to apply for a plan of work for that area provided it offers in good faith to include the Enterprise as a joint-venture partner".[72]

***

Despite the reforms introduced by the 1994 Agreement in Part XI, which favor the industrialized States, it is notable that the fundamental elements of the international legal framework governing the ocean seabed remain intact. The Area and its resources are to be regarded as "the common heritage of mankind", as expressed through the principle of non-appropriation. Consequently, activities conducted within the Area must be undertaken for the benefit of mankind as a whole. The "parallel system" also remains in place, according to which States and their nationals can have access to seabed mineral resources, in connection with the existence of reserved areas, as described in detail above.[73]

Moreover, the current international regime for the exploitation of the international seabed, as described here, is an integrated model for resource management and exploitation. While in space the relevant discussions and considerations are still at the (embryonic) level of formulating principles (in particular, whether the principle of non-appropriation will be maintained or modified), in the maritime domain the international regime constituted by Part XI of UNCLOS, as amended and currently in force under the 1994 Agreement, constitutes a comprehensive and detailed resource exploitation regime. A regime which is clearly international in character, which has a sophisticated institutional structure (the Authority, its organs, their interaction with States), which provides for the rights and obligations of all types of actors in the Area, which clearly defines the procedures and actions to be taken there, and which has defined in detail the resources sought. In its current operational phase, this regime will be further refined by adopting the above-mentioned Regulations. Regardless of the degree of acceptance (ideological or operational/business) of this regime of exploitation of the international seabed, regardless of the fact that this regime is only now being put into effect, regardless of what benefits (for the Authority, for the States) will ultimately emerge from its operation, the existence of this regime is a constant reminder of how much remains to be done in outer space if the international community really intends to exploit space resources.

However, it must be stressed that the concept of the common heritage of mankind (CHM) is intrinsically contradictory,[74] in that it seeks to reconcile the imperatives of development and exploitation with the need to protect the marine environment. This fundamental antinomy remains important, with more recent environmental

[72] 1994 Agreement, Section 2, para. 5.

[73] See Tanaka, *op. cit.*, p. 231–232; UNCLOS, Article 153.

[74] A. Jaeckel, "Benefitting from the Common Heritage of Humankind: From Expectation to Reality", The International Journal of Marine and Coastal Law, Vol. 35(4), 2020, p. 3.

considerations tending to completely reverse the balance.[75] In any event, the exploitation of the resources of the international seabed should be reconsidered in the light of the imperatives of sustainable development. This is also crucial for any future consideration of the legal regime for the exploration and exploitation of space resources.[76] It should be noted, however, that the concept of "environmental protection" is not raised with the same intensity at the terrestrial and extraterrestrial levels. Irrespective of the depth, the international seabed resources and the oceanic environment in general are located on Earth, and consequently there is a greater sense of proximity compared to the environment of celestial bodies, which are "far away", at least according to the current anthropocentric spirit that dominates space law and the psychology of States as far as the space adventure is concerned. It should also be remembered that the need to avoid so-called forward contamination in outer space exists mainly for scientific reasons and much less because mankind feels the need not to harm the environment of celestial bodies. Thus, the importance of planetary protection and the consequent setting of limits on the exploitation of space resources will depend on humanity's interest in protecting the space (extraterrestrial) environment, for which there may not be the same sensitivity that has existed—and recently intensified—for the Earth's marine environment and the international seabed. The anthropocentric rationale of space law—especially of the Outer Space Treaty—undoubtedly signals a greater interest in 'back contamination' than in 'forward contamination'—at least until now.

In view of this, it is likely that the need to protect the space environment (in the forward contamination dimension) will not be an obstacle to the exploitation of space resources, at least to the extent that this need has already been expressed with respect to the Earth's oceans. Otherwise, however, the current international regime for the exploitation of the international seabed, despite its complexity and possible shortcomings, constitutes a regulatory framework that will *de facto* influence the legal regime for the exploitation of space resources that is currently being developed. A decisive factor in the orientation of the latter will be, *inter alia*, the establishment of a regime for the equitable sharing of the benefits from the exploitation of resources within the framework of UNCLOS and the 1994 Agreement, as discussed below.

[75] An interesting environmental issue that has now emerged is protecting marine genetic resources found in hydrothermal vents. Ocean mining may be crucial to the conservation of these resources, which are of great scientific and commercial interest. It should be noted that the 2023 Agreement on Marine Biodiversity of Areas beyond National Jurisdiction (BBNJ Agreement) establishes a prior (and post-) notification mechanism for the collection of marine genetic resources *in situ* in areas beyond national jurisdiction (Article 12). – See, in this respect, S. Lothian, "How does the new BBNJ Agreement regulate access to and benefit-sharing of marine genetic resources at deep-sea hydrothermal vents?", 1st June 2023, ECO LAW Blog, https://blogs.ed.ac.uk/eco-law/2023/06/01/how-does-the-new-bbnj-agreement-regulate-access-to-and-benefit-sharing-of-marine-genetic-resources-at-deep-sea-hydrothermal-vents/ (Last accessed: 26.1.2025).

[76] See also *supra*, under II.1.1. c.

6.1.2.3 Sharing the Benefits

Equitable sharing of benefits is clearly defined in the relevant international instruments with regard to international seabed resources, while it is also required in the context of other important international agreements. The question is whether it can be argued that a similar obligation exists in outer space, particularly with respect to space resources.

6.1.2.3.1 Equitable Benefit Sharing of the International Seabed and Marine Resources

As a direct consequence of the concept of the "common heritage of mankind", the essence of the international legal regime governing the exploitation of the international seabed is twofold: on the one hand, all rights to the resources of the Area are vested in mankind as a whole and the relevant activities in the Area are carried out for its benefit.[77] On the other hand, the principle of non-appropriation applies (as in outer space): "No State shall claim or exercise sovereignty or sovereign rights over any part of the Area or its resources, nor shall any State or natural or juridical person appropriate any part thereof".[78] Consequently, this regime establishes first a "benefit of mankind". This is, after all, the title of Article 140 UNCLOS. Obviously, this benefit "of mankind" extends to the resources of the Area: "All rights in the resources of the Area are vested in mankind as a whole, on whose behalf the Authority shall act. These resources are not subject to alienation".[79] It can be argued that this legal construction allows (if not requires) the consideration of seabed resources as collective goods (of Humanity).[80]

However, this "benefit to mankind" is not absolute, as it can be shared according to the UNCLOS. Article 137 provides that "minerals recovered from the Area… may… be alienated in accordance with this Part and the rules, regulations and procedures of the Authority", which shall provide for "the equitable sharing of financial and other economic benefits derived from activities in the Area through any appropriate mechanism, on a non-discriminatory basis".[81] The rules, regulations and procedures on the equitable sharing are considered and approved by the Assembly.[82]

Article 140 refers to "financial and other economic benefits" and to an "appropriate mechanism" for sharing these benefits. Although it is reasonable to assume that non-financial benefits might include, *inter alia*, the transfer of relevant technology,

---

[77] Arts. 137, 140 UNCLOS.

[78] Art. 137 para. 1 UNCLOS.

[79] Art. 137 para. 2 UNCLOS.

[80] See D. Wilde, H. Lily, N. Craik & A. Chakraborty, "Equitable sharing of deep-sea mining benefits: More questions than answers", Marine Policy, Vol. 151, 2023, p. 2.

[81] Art. 140 para. 2 UNCLOS.

[82] Art. 160, para. 2(f)(i) UNCLOS.

potential environmental benefits (e.g. reduced environmental impact compared to land-based mining) or even an increase in biodiversity through the discovery of new forms of marine life,[83] the Convention does not appear to make any explicit reference to this. The same can be said of the "mechanism" through which sharing is to take place. It is therefore reasonable to assume that the main benefit, at least on the basis of the information available so far, will be the sharing of income generated by the extraction of mineral resources in the Area.[84]

As said, according to the 1994 Agreement, rules, regulations and procedures on the equitable sharing of financial and other economic benefits derived from activities in the Area and the relevant decisions shall be based on recommendations of the Finance Committee.[85] In July 2023, the said committee considered a draft proposal for the establishment of a "Common Heritage Fund",[86] in the context of implementing the 2030 Agenda for Sustainable Development and achieving the Sustainable Development Goals.[87] This fund should have the potential to facilitate the direct distribution of financial benefits, and also to support global public goals, investment in human and physical capital or deep-sea research.[88]

In practice, though, the actual potential economic benefit of exploiting the mineral resources of the international seabed is still unclear, without of course considering the environmental constraints of such exploitation.[89] As said, Article 145 UNCLOS provides for "necessary measures" with respect to activities in the Area "to ensure effective protection for the marine environment from harmful effects which may arise from such activities". In this respect, the Authority must adopt "appropriate rules, regulations and procedures". Protecting the marine environment is even more important today. Peter Thomson, UN Secretary-General's Special Envoy for the Ocean, recently pointed out that "the Ocean plays a fundamental planetary role in mitigating the effects of Climate Change through its absorption of heat and carbon. This service alone makes the healthy functioning of the Ocean's ecosystems critical to all humans".[90] And it was Thomson who called for a 10-year moratorium with respect to deep-sea mining in 2019.[91]

---

[83] See https://deepseamining.ac/support#gsc.tab=0 (last accessed: 03.03.2025).

[84] Wilde et al., *op. cit.*, p. 2.

[85] 1994 Agreement, Section 9, para. 7 (f).

[86] ISBA/28/FC/4.

[87] See https://www.isa.org.jm/equitable-sharing-of-benefits/ (last accessed: 31.01.2025).

[88] Idem.

[89] See Jaeckel, *op. cit.*, p. 6 and 17.

[90] A message from Ambassador Peter Thomson, UNSG's Special Envoy for the Ocean to the IOC-UNESCO Ocean Literacy Training Courses, 19.1.2025, https://sdgs.un.org/news/message-ambassador-peter-thomson-unsgs-special-envoy-ocean-ioc-unesco-ocean-literacy-training (last visit: 19.1.2025).

[91] Julián Reingold, "Will Deep Sea Mining Suffocate Ocean Conservation?", April 11, 2024, Earth Island Journal, https://www.earthisland.org/journal/index.php/articles/entry/deep-sea-mining-is-suffocating-ocean-conservation/## (last accessed: 19.01.2025).

This uncertainty primarily impacts the expectations of developing countries, especially when taking into account the potential repercussions of mining in the Area on those (in most cases, developing) countries that are land-based producers of such resources. The UNCLOS provided for a “system of compensation” or “other measures of economic adjustment assistance” to assist those developing countries that would be economically affected by the extraction of mineral resources in the Area, because they are exporters of those minerals (UNCLOS, Art. 151 para. 10), on the basis of a proposal from the Economic Planning Commission to the Authority’s Council [UNCLOS, Art. 164 para. 2 (d)]. In accordance with the 1994 Agreement, this aid will come from an economic assistance fund that is established, “from a portion of the funds of the Authority which exceeds those necessary to cover the administrative expenses of the Authority” (1994 Agreement, Section 7). Further, “The rates of payments under the system shall be within the range of those prevailing in respect of land-based mining of the same or similar minerals in order to avoid giving deep seabed miners an artificial competitive advantage or imposing on them a competitive disadvantage” (1994 Agreement, Section 8).[92]

Transfer of technology. Regarding the transfer of technology in connection with activities in the Area (which also represents a benefit to be shared), Article 144 of UNCLOS provided for the facilitation of access by both the Enterprise and developing States to the relevant technology “under fair and reasonable terms and conditions”. However, the 1994 Agreement completely reversed the above regime, since, under Section 5 of the Annex to that Agreement, the relevant technology is now to be sought “on fair and reasonable commercial terms and conditions on the open market, or through joint-venture arrangements”, while the Parties have only the ‘obligation’ to “facilitate the acquisition” of such technology, “consistent with the effective protection of intellectual property rights”. As a general principle, there is a ‘neutral’ obligation to “promote international technical and scientific cooperation with regard to activities in the Area”. These changes make the ultimate benefits of technology transfer (for developing countries) uncertain.

Benefit sharing in other international agreements. Although it is still largely undefined, the equitable sharing of benefits is a clear option in other relevant international instruments. The 1992 Convention on Biological Diversity (CBD)[93] has as its main purpose—in addition to the conservation of biological diversity and the sustainable use of its components—“the fair and equitable sharing of the benefits arising out of the utilization of genetic resources, including by appropriate access to

[92] The issue of remuneration for land-based producers has been of particular concern to the African Group. This concern has been expressed on multiple occasions (see, for instance, African Group Submission on the ISA Payment Regime for Deep-Sea Mining in the Area, 5.7.2019, https://www.isa.org.jm/wp-content/uploads/2022/06/agsmitmodelfinal.pdf, and June 2022, https://www.isa.org.jm/wp-content/uploads/2022/12/African_Group_Submission_Payment_Regime.pdf). For these issues, also to see Wilde, et al., *op. cit.*, p. 9.

[93] Convention on Biological Diversity, adopted on 5 June 1992 (Rio de Janeiro), entered into force on 29 December 1993, 1760/U.N.T.S./79, 31 I.L.M. 818 (1992).

genetic resources and by appropriate transfer of relevant technologies" (Article 1). The 2010 Nagoya Protocol,[94] supplementary to the CBD, provides that "benefits arising from the utilization of genetic resources as well as subsequent applications and commercialization shall be shared in a fair and equitable way with the Party providing such resources that is the country of origin of such resources or a Party that has acquired the genetic resources in accordance with the Convention. Such sharing shall be upon mutually agreed terms" (Article 5).

In addition, the 2023 Agreement on Marine Biological Diversity of Areas beyond National Jurisdiction (BBNJ Agreement)[95] provides for the fair and equitable sharing of benefits arising from activities with respect to marine genetic resources of areas beyond national jurisdiction while ensuring the conservation and sustainable use of marine biological diversity of the areas concerned (Art. 14). The same provision makes a distinction between monetary benefits and non-monetary benefits, with the latter including, but not limited to, access to samples and sample collections, access to digital sequence information, open access to scientific data, transfer of marine technology, capacity-building, and increased technical and scientific cooperation (in particular with scientists and institutions in developing States). Monetary benefits shall be shared "fairly and equitably", through the financial mechanism established under the Agreement.

6.1.2.3.2 Sharing the Benefits of Outer Space?

In theory, it has been argued that a benefit-sharing regime would seem to follow from the wording of Art. I para. 1 of the OST.[96] However, a careful reading of this provision would point out that it is not the space activities of "each State" that "shall be carried out for the benefit and in the interests of all countries", but rather "the exploration and use of outer space" in general. As such, the provision is mainly declaratory in nature, emphasizing that space is a common field of action for all States,[97] rather than establishing an explicit obligation to share the benefits of this

[94] Nagoya Protocol on Access to Genetic Resources and the Fair and Equitable Sharing of Benefits Arising from their Utilization to the Convention on Biological Diversity, adopted on 29 October 2010, entered into force on 12 October 2014, 3008/U.N.T.S./3.

[95] Agreement under the United Nations Convention on the Law of the Sea on the Conservation and Sustainable Use of Marine Biological Diversity of Areas beyond National Jurisdiction, New York, 19 June 2023, open for signature on 20 September 2023–until 20 September 2025. Not yet in force.

[96] R. Deplano, "Inclusive Space Law: The Concept of Benefit Sharing in the Outer Space Treaty", I.C.L.Q., Vol. 72, July 2023, p. 677 ("Article I, paragraph 1 is the only provision of the Outer Space Treaty which deals with the obligation of benefit sharing...").

[97] In the absence of any right of appropriation, as guaranteed, moreover, by Article II of the Outer Space Treaty.

action.[98] Nor does the so-called Space Benefits Declaration[99] seem to shed more light on Article 1 para. 1 OST, as it focuses mainly on strengthening cooperation between States "for the mutual benefit and in the interest of all parties involved". Such cooperation should be defined "on an equitable and mutually acceptable basis", and the relevant contractual terms "should be fair and reasonable".

Throughout the Declaration, the concept of benefit-sharing is only implicitly addressed in the context of scientific and technological advancement and capacity building, where it is stated, in rather general terms, that:

> International cooperation, while taking into particular account the needs of developing countries, should aim, *inter alia*, at the following goals, considering their need for technical assistance and rational and efficient allocation of financial and technical resources:

(a) Promoting the development of space science and technology and of its applications;
(b) Fostering the development of relevant and appropriate space capabilities in interested States;
(c) Facilitating the exchange of expertise and technology among States on a mutually acceptable basis.

This resolution serves mainly as a call to States, since its language carefully avoids prescribing any specific conduct.[100] Nevertheless, it can be argued that this general appeal (to promote international cooperation) is also applicable to the more specific area of space resource exploitation.

In contrast, the Moon Agreement explicitly provides that the international exploitation regime under (its) Article 11 will be a benefit-sharing regime, since paragraph 7 of the same provision, which refers to the main purposes of this regime, clearly

[98] As rightfully mentioned, "Benefit sharing is not clearly defined or regulated in the existing international space law framework, which creates legal uncertainty and challenges for space activities such as space resource utilization" - Grant Henriksen, "New NSS Paper on Sharing the Benefits of Space", October 3, 2023, https://nss.org/new-nss-paper-on-sharing-the-benefits-of-space/ (last visited: 8.12.2024). See also Bohacek et al., who note that "The precise meaning of the OST's Article I benefit-sharing clause for SRU remains a topic of interpretations, and even by rejection…Particularly contested is the question of what the obligations to share the benefits of outer space activities are" – P. Bohacek, S.P. Worden & K. Grattan, "Benefit-Sharing as Investment Protection for Space Resource Utilization", New Space, Volume 10, Number 2, 2022, https://doi.org/10.1089/space.2021.0050. Also, F. Xu & J. Su, who admit that "Without concrete meaning of the common benefit and interests principle, abiding by the obligation to share benefits might have to depend largely on the good faith of States" - "Towards a legal regime of benefits sharing for space mining: With some experience from the Area", Resources Policy, Volume 76, 2022, article id. 102627, https://doi.org/10.1016/j.resourpol.2022.102627.

[99] A/RES/51/122, Declaration on International Cooperation in the Exploration and Use of Outer Space for the Benefit and in the Interest of All States, Taking into Particular Account the Needs of Developing Countries, General Assembly, 51st session, 13.12.1996.

[100] "States are free to determine all aspects of their participation in international cooperation…"; "All States, particularly those with relevant space capabilities and with programmes for the exploration and use of outer space, should contribute to promoting and fostering international cooperation on an equitable and mutually acceptable basis…".

emphasizes that it should include, *inter alia*, “an *equitable sharing by all States Parties in the benefits* derived from [the] resources”. And it is characteristic that, always in accordance with the same paragraph, this sharing is clearly guided by “the interests and needs of the developing countries, as well as the efforts of those countries which have contributed either directly or indirectly to the exploration of the moon”.[101] This is, of course, consistent with the proclamation in the Moon Agreement that the Moon is the common heritage of mankind, which, in the language of the Agreement, “finds its expression” in Article 11 para. 5 and the establishment of an international exploitation regime. In this sense, the exploitation regime of the Moon Agreement is much more targeted than the general approach of the Outer Space Treaty and, under the state of the “common heritage of mankind”, considers the exploitation of the resources of the Moon (and other celestial bodies within the Solar System, according to the explicit reference of Article 1 para. 1) as the common property of the global community. Thus, the exploitation regime of the Moon Agreement has a specific character and orientation with respect to the generic, principle-level proclamations of the Outer Space Treaty and, under the concept of the “Common Heritage of Mankind”, treats the resources of the Moon (and other celestial bodies within the Solar System)[102] as the common property of the global community.

Although the Moon Agreement has not yet been widely ratified by States, it must be accepted that its provision for the equitable sharing of the benefits arising from space resources is technically in line with what is established in other international instruments of a similar nature, such as the international instruments on deep seabed resources and marine genetic resources referred to above.

Thus, despite its lack of widespread acceptance, the Moon Agreement seems to follow an already established trend in international law: the provision, through international arrangements, for the equitable sharing of benefits in relation to resources that are either beyond national jurisdiction or of particular importance to humanity as a whole. Leaving aside the specific parameters under which this sharing is (or can be) implemented in each case, and even the possible gaps, the trend is clear. As such, the equitable sharing of benefits, combined with the establishment of an international legal regime, cannot be overlooked as a valid model for regulating the exploitation of space resources.

6.1.3 Re-positioning International Law: Emergence of New Customary Rules

Given that custom is a recognized source of international space law—and that, in certain circumstances, it may even form rapidly or “instantaneously”—the prospective proliferation of national laws concerning the exploration and use of space

[101] Moon Agreement, Art. 11 para. 7 (d).

[102] … in accordance with the explicit reference in Article 1, para. 1 MOON.

resources could, under specific conditions, contribute to the emergence of new rules at the customary level.

6.1.3.1 Custom as a Source of International Space Law: The Customary Dimension of the Fundamental Principles

As already mentioned, the origins of international space law can be traced back to a series of resolutions of the United Nations General Assembly, adopted during the decade between the launch of the first Sputnik (1957) and the adoption of the Treaty on Outer Space (1967):

1148 (XXI) of 14 November 1957 (Regulation, limitation and balanced reduction of all armed forces and all armaments; Conclusion of an international convention (treaty) on the reduction of armaments and the prohibition of atomic, hydrogen and other weapons of mass destruction), which, *inter alia*, states: "Urges that the States concerned. ... give priority to reaching a disarmament agreement which, upon its entry into force, will provide for the following: ... (f) the joint study of an inspection system designed to ensure that the sending of objects through outer space shall be exclusively for peaceful and scientific purposes;".

1348 (XIII) of 13 December 1958 (Question of the Peaceful Use of Outer Space), which established the Committee on the Peaceful Uses of Outer Space, and by:

1472 (XIV) of 12 December 1959 (International Co-operation in the Peaceful Uses of Outer Space), creating the Committee on the Peaceful Uses of Outer Space (UN COPUOS) (still in existence today).

1721 (XVI) of 20 December 1961 (International Co-operation in the Peaceful Uses of Outer Space), in which the principles of freedom of exploration and use of outer space and the application of international law, including the Charter, to outer space are established for the first time (later Articles I and III OST): "(a) International law, including the Chapter of the United Nations, applies to outer space and celestial bodies; (b) Outer space and celestial bodies are free for exploration and use by all States in conformity with international law and are not subject to national appropriation;".

1884 (XVIII) of 17 October 1963 (Question of general and complete disarmament), which calls upon States "(a) to refrain from placing in orbit around the earth any objects carrying nuclear weapons of mass destruction, installing such weapons on celestial bodies, or stationing such weapons in outer space in any other manner;" (subsequently Article IV(1) OST).

1962 (XVIII) of 13 December 1963 (Declaration of Legal Principles Governing the Activities of States in the Exploration and Use of Outer Space), in which the General Assembly 'solemnly declares' the fundamental principles of the law of outer space, 4 years before they were codified in the 1967 Outer Space Treaty:

1. The exploration and use of outer space shall be carried on for the benefit and in the interests of all mankind.

2. Outer space and celestial bodies are free for exploration and use by all States on a basis of equality and in accordance with international law.
3. Outer space and celestial bodies are not subject to national appropriation by claim of sovereignty, by means of use or occupation, or by any other means.
4. The activities of States in the exploration and use of outer space shall be carried on in accordance with international law, including the Charter of the United Nations, in the interest of maintaining international peace and security and promoting international cooperation and understanding.
5. States bear international responsibility for national activities in outer space, whether carried on by governmental agencies or by non-governmental entities, and for assuring that national activities are carried on in conformity with the principles set forth in the present Declaration. The activities of non-governmental entities in outer space shall require authorization and continuing supervision by the State concerned. When activities are carried on in outer space by an international organization, responsibility for compliance with the principles set forth in this Declaration shall be borne by the international organization and by the States participating in it.
6. In the exploration and use of outer space, States shall be guided by the principle of cooperation and mutual assistance and shall conduct all their activities in outer space with due regard for the corresponding interests of other States. If a State has reason to believe that an outer space activity or experiment planned by it or its nationals would cause potentially harmful interference with activities of other States in the peaceful exploration and use of outer space, it shall undertake appropriate international consultations before proceeding with any such activity or experiment. A State which has reason to believe that an outer space activity or experiment planned by another State would cause potentially harmful interference with activities in the peaceful exploration and use of outer space may request consultation concerning the activity or experiment.
7. The State on whose registry an object launched into outer space is carried shall retain jurisdiction and control over such object, and any personnel thereon, while in outer space. Ownership of objects launched into outer space, and of their component parts, is not affected by their passage through outer space or by their return to the Earth. Such objects or component parts found beyond the limits of the State of registry shall be returned to that State, which shall furnish identifying data upon request prior to return.
8. Each State which launches or procures the launching of an object into outer space, and each State from whose territory or facility an object is launched, is internationally liable for damage to a foreign State or to its natural or juridical persons by such object or its component parts on the Earth, in air space, or in outer space.
9. States shall regard astronauts as envoys of mankind in outer space, and shall render to them all possible assistance in the event of accident, distress, or

> emergency landing on the territory of a foreign State or on the high seas. Astronauts who make such a landing shall be safely and promptly returned to the State of registry of their space vehicle.

The above-mentioned UN General Assembly Resolutions reflect the practice of the States that were actively involved in space affairs in that "early" age of space activities (USA–USSR), but also, more generally, the positions of the international community as a whole. As already mentioned, it was during this period that the principles that would be codified in their final form in the *Magna Carta* of international space law, the Outer Space Treaty of 1967, began to take shape, principles that were conceived in a very short time through the practice of the above-mentioned States in the exploration and use of outer space. This point is emphasized in particular by Vereshchetin and Danilenko, who, while advocating the primacy of conventional space law over customary law, nevertheless recognize that custom has historically been the first source of such law and that the practice of States in the field of exploration and use of outer space has been instrumental in the emergence of important principles and rules of international space law.[103]

The "instant" formation of customary space law. A particular feature of the formation of customary norms in the field of international space law is the speed with which they are created. As Professor Emmanuel Roucounas observes, "in some cases, in order to anticipate developments, the international community 'burns the time factor'".[104] As already pointed out, Declaration 1962 (XVIII), which provides the first systematic classification of the fundamental principles of international space law, was adopted by the UN General Assembly only 6 years after the beginning of space activities (1957).

Furthermore, the placement of artificial satellites in orbit led to the creation of another 'instant' customary rule, according to which States had (acquired) the right to transit the airspace of other States during the process of ascent of the rockets carrying the satellites to the orbital plane. Obviously, this rule was based on the practice of the lack of protest from the states concerned regarding the transit of space objects.

Vereshchetin & Danilenko attribute the rapid formation of customary international space law to the fact that all activities and developments related to the exploration and use of outer space became immediately known at the global level, as well as to the continuous involvement of the UN General Assembly in this regard, with the adoption of a number of resolutions.[105] For his part, Bin Cheng referred to the "one-element custom", i.e. the practice of States, since the practice itself plays an

[103] V.S. Vereshchetin & G.M. Danilenko, «Custom as a source of International Law of Outer Space», J. Space L., Vol. 13, No 1, 1985, p. 25.

[104] Emm. Roucounas, Public International Law, 3rd Edition, 2019, Nomiki Vivliothiki, p. 57 (In Greek).

[105] V.S. Vereshchetin & G.M. Danilenko, «Custom as a source of International Law of Outer Space», J. Space L., Vol. 13, No 1, 1985, p. 25.

evidentiary role in the establishment of the custom, the formation of which becomes "instant international customary law".[106]

Moreover, the link between the resolutions of the UN General Assembly and the formation of an international custom has also been accepted by the International Court of Justice. In the case of *Military and Paramilitary Activities in and against Nicaragua* (1986), the Court held that the acceptance by States of the text of a number of important General Assembly resolutions should be regarded as an acceptance of the rules contained in the resolution in question.[107] Furthermore, in its Opinion on the *Legality of the Threat or Use of Nuclear Weapons* (1996), the ICJ accepted that the repetition of General Assembly resolutions may entail "the gradual evolution of the *opinio juris* necessary for the establishment of a new rule".[108]

6.1.3.2 The Impact of National Laws on the Emergence of International Space Law Customary Rules

The adoption by certain States of national laws relating to the exploitation of the mineral resources of celestial bodies, which effectively grant property rights over such resources, is in apparent contradiction to the principle of non-appropriation (Article II OST), at least according to the currently prevailing view. From a legal point of view, significant legal consequences may arise if more States follow this practice by adopting domestic legislation to encourage the commercial extraction of resources from planets and asteroids by non-governmental entities.[109] If such a practice were to be followed by States with significant activity in outer space, possibly a major institutional change (of practice) would take place at the level of international space custom. This change could potentially lead to a new interpretative approach to the principle of non-appropriation [enshrined at both the customary and conventional level (Article II OST)], which, in practice, would amount to a modification of the provision through (new) practice. Such a process is perfectly conceivable in the context of international law, since there is no established hierarchy between its main sources (treaty–custom) and, therefore, a new customary rule is capable of modifying, even abolishing, a conventional (but, obviously, also a customary) rule.[110]

If such a scenario were to materialize with regard to the extraction of space resources, a new but legally acceptable international regime for the exploitation of space resources could emerge, but it would be a source of great legal uncertainty: the interaction between the States that would commit to and implement the new

---

[106] Bin Cheng, Studies in International Space Law, Clarendon Press, Oxford, 1997, p. 137.

[107] ICJ, Militarv and Paramilitary Activities in and against Nicaragua (Nicaragua v. United States of America), Merits, Judgment. I.C.J. Reports 1986, p. 14, at para.188.

[108] ICJ, Legality of the Threat or Use of Nuclear Weapons, Advisory Opinion, I.C.J. Reports 1996, p. 226, at para. 70.

[109] For the existing national legislations that provide regulation for space resources, see *supra*, I.2.2.

[110] P.-M. Dupuy & Y. Kerbrat, Droit International Public, 13e ed., 2016, p. 298.

(customary) regime, those that would continue to accept the currently prevailing (highly restrictive) interpretation of the principle of non-appropriation, and those that have ratified the 1979 Moon Agreement would constitute uncharted waters for international law.

6.2 Bypassing the Gordian Knot: Consulting the International Court of Justice

In addition to its power to adjudicate on disputes between States, the International Court of Justice has the power to give advisory opinions. These are two distinct functions of the Court,[111] both in terms of the content of the Court's rulings and their addressees.

The jurisdiction of the Court to settle disputes between States is provided for in its Statute. According to Article 34 of the ICJ Statute, only States may be parties in cases before the Court. Further, the Court has jurisdiction to adjudicate "all cases which the parties refer to it and all matters specially provided for in the Charter of the United Nations or in treaties and conventions in force" (Article 36 para. 1). States may at any time make a declaration recognizing as compulsory the jurisdiction of the Court, "*ipso facto* and without special agreement, in relation to any other state accepting the same obligation" (Article 36 para. 2). Pursuant the same provision, disputes before the Court may concern: "the interpretation of a treaty; any question of international law; the existence of any fact which, if established, would constitute a breach of an international obligation; the nature or extent of the reparation to be made for the breach of an international obligation". In this respect, the applicable law (for the settlement of the legal disputes) is defined in Article 38 para. 1 of the Statute and includes: "international conventions, whether general or particular, establishing rules expressly recognized by the contesting states; international custom, as evidence of a general practice accepted as law; the general principles of law recognized by civilized nations;". Judicial decisions and the teachings "of the most highly qualified publicists of the various nations" are considered "as subsidiary means for the determination of rules of law".

In contrast, the advisory function of the Court is provided for in Article 96 of the Charter of the United Nations. According to this provision,

UN Charter, Article 96

1. The General Assembly or the Security Council may request the International Court of Justice to give an advisory opinion on any legal question.

[111] As the Court itself made clear, "The purpose of the advisory function is not to settle - at least directly - disputes between States, but to offer legal advice to the organs and institutions requesting the opinion" - Legality of the Threat or Use of Nuclear Weapons, Advisory Opinion, I.C.J. Reports 1996, p. 226, at para. 15.

2. Other organs of the United Nations and specialized agencies, which may at any time be so authorized by the General Assembly, may also request advisory opinions of the Court on legal questions arising within the scope of their activities.

It follows that the advisory function of the Court is not accessible to States (contrary to its competence to settle international disputes) but is in principle open to the General Assembly and the Security Council on any matter on which these organs consider that they wish to submit a question to the Court. In addition, other organs of the United Nations (e.g. ECOSOC) as well as international intergovernmental organisations, which are specialised agencies of the United Nations system, may request an advisory opinion, subject to two restrictions: the approval of the UN General Assembly is required and the question must fall within the scope of the organ or international organization.[112] The scope of Article 96 can be justified by the fact that the International Court of Justice is an organ of the United Nations. The Court's predecessor, the Permanent Court of International Justice, had a similar competence.[113]

Furthermore, Article 65 of the ICJ Statute provides that the Court has a discretion—not an obligation—to give an opinion and describes in detail how the question is to be submitted to the Court:

ADVISORY OPINIONS
Article 65

1. The Court may give an advisory opinion on any legal question at the request of whatever body may be authorized by or in accordance with the Charter of the United Nations to make such a request.
2. Questions upon which the advisory opinion of the Court is asked shall be laid before the Court by means of a written request containing an exact statement of the question upon which an opinion is required, and accompanied by all documents likely to throw light upon the question.

However, it should be clarified that the Court has never refused to give an opinion when it has been duly consulted.[114] The Court's opinions are not binding, unless it is expressly specified, as in the case of the—aforementioned—1946 Convention on the Privileges and Immunities of the United Nations.

Since 1948, the International Court of Justice has issued 28 advisory opinions. Some of its opinions have been criticized, such as, for example, the opinions on the

[112] It is interesting to note that Article VIII, section 30, of the Convention on the Privileges and Immunities of the United Nations, Adopted by the General Assembly of the United Nations on 13 February 1946, provides, *inter alia*, that "…If a difference arises between the United Nations on the one hand and a Member on the other hand, a request shall be made for an advisory opinion on any legal question involved in accordance with Article 96 of the Charter and Article 65 of the Statute of the Court. The opinion given by the Court shall be accepted as decisive by the parties".

[113] See Roucounas, *op. cit.*, p. 542.

[114] Roucounas, *op. cit.*, p. 544.

"*Accordance with international law of the unilateral declaration of independence in respect of Kosovo*", on the "*Legal Consequences of the Construction of a Wall in the Occupied Palestinian Territory*"[115] or on the "*Legality of the Threat or Use of Nuclear Weapons*".[116] In other cases, however, the opinions delivered by the Court have decisively clarified outstanding questions in international relations and have contributed significantly to the progressive development of international law in this field. Reference may be made here to the now historic advisory opinions: "*Conditions of Admission of a State to Membership of the United Nations (Article 4 of the Charter)*" of 28 May 1948,[117] and "*Certain Expenses of the United Nations (Article 17(2) of the Charter)*" of 20 July 1962,[118] which clarified important questions concerning the legal personality and general functioning of international organisations and had a catalytic effect on their recognition as autonomous subjects of international law, possessing "a will of their own" distinct from that of the Member States.

***

In addition to any criticism of the advisory jurisdiction of the International Court of Justice, it is accepted that this function "may and do develop the law".[119] This conclusion is very useful should the international community decide to involve the International Court of Justice in the issue of space resources and seek its opinion, as this book argues.

An advisory opinion for space resource activities. Throughout this book, the impasse in which the international space community finds itself on the issue of regulating space resource activities has been highlighted. This impasse becomes apparent when one considers that, on the one hand, everyone seems to agree on the need for such activities to take place, and no one seems to deny the participation and involvement of private entities in space affairs, but, on the other hand, there is a general disagreement—which, as has been shown, manifests itself in a variety of ways—as to what legal regime would be appropriate for the effective regulation of the activity in question. Add to this the absolute reluctance, at least at present, of a significant number of States to accept a new international treaty on space activities, and the above-mentioned impasse is exacerbated. Moreover, all the alternative approaches that have been proposed—beyond the conventional creation of an international legal framework—namely the gradual formation of relevant customary

---

[115] See, for instance, Joel S. Tashjian, "Contentious Matters and the Advisory Power: The ICJ and Israel's Wall", Chicago J.I.L., Vol. 6 No 1, 6.1.2005, Article 25, p. 427–437.

[116] Accordance with International Law of the Unilateral Declaration of Independence in Respect of Kosovo, Advisory Opinion, I.C.J. Reports 2010, p. 403; Legal Consequences cf. the Construction of a Wall in the Occupied Palestinian Territory, Advisory Opinion, I.C.J. Reports 2004, p. 136; Legality of the Threat or Use of Nuclear Weapons, Advisory Opinion, I.C.J. Reports 1996, p. 226.

[117] I.C.J. Reports 1948, p. 57.

[118] I.C.J. Reports 1962, p. 151.

[119] Teresa F. Mayr/Jelka Mayr-Singer, "Keep the Wheels Spinning: The Contributions of Advisory Opinions of the International Court of Justice to the Development of International Law", ZaöRV 76 (2016), p. 425–449, at p.426.

rules or the emergence of a set of non-binding principles, can only function as provisional substitutes, since, as already mentioned, they will ultimately be ineffective in the long run.

Faced with this impasse, it is now suggested that the International Court of Justice be asked to provide an advisory opinion on the issue of space resources.

Technically, if the States agree, recourse to the Court is feasible, since UNCOPUOS is a subsidiary body of the General Assembly. Thus, either COPUOS could submit the relevant question—or questions—to the Court after the General Assembly has given its approval, or the latter could submit the questions directly to the Court on the recommendation of COPUOS.

What questions might be put to the Court? There could be many suggestions, the common feature of which would be that the questions should be formulated in such a way as to enable the Court to give a substantive opinion which would help to formulate a generally acceptable legal framework. In any event, the questions could take the following form:

- **Does the exploration and exploitation of mineral resources of the celestial bodies constitute a violation of the principle of non-appropriation enshrined in Article 2 of the Treaty on Principles Governing the Activities of States in the Exploration and Use of Outer Space, including the Moon and Other Celestial Bodies?**
- **Furthermore, what are the applicable rules of international law under which the exploration and exploitation of mineral resources of the celestial bodies must take place?**

In any case, the advisory opinion of the International Court of Justice will be decisive for the development of the regulation of these activities. If the Court answers the first question in the negative and the second question in the affirmative, its opinion will constitute an interpretation of the Outer Space Treaty that can be universally accepted, despite the fact that advisory opinions are not binding. If the answer to the first question is in the affirmative, the international community as a whole will now be aware that, in order to effectively regulate the use of space resources, it is necessary to set in motion an international law-making process which, if it does not go so far as to amend the existing provisions of space law, will at least seek to establish an international regime of exploitation acceptable under existing law.

A possible recourse, as described above, to the advisory function of the International Court of Justice will help the international space community to break the legal deadlock it is currently experiencing on this issue. And it will undoubtedly help international law to finally take the desired decisive step towards a substantive regulation of the use and exploitation of space resources, away from hesitations which, instead of being beneficial, ultimately cancel out activities that are now desired by the whole of humankind.

***

The necessary revival of international law concerning the exploration and use of space resources may be pursued through either a radical approach or a more modest, "administrative" drafting of non-binding principles. The "hard" option would entail an agreement among the international community to adopt new conventional rules (either *ad hoc* or as part of a comprehensive reformulation of space law), which could potentially result in a transgression of established principles of the existing legal framework. On the other hand, States may opt for less intrusive processes, such as the gradual emergence of customary rules grounded in state practice, which would be reinforced by the (possible) expansion of national legislative approaches to the subject.

Nevertheless, the current legal regime governing the exploration and exploitation of international seabed resources, though controversial and still in its early operational stages, remains the only comprehensive, detailed, and coherent legal model that could inspire a similar framework for space resource exploration and utilization. Key characteristics of this model include its international nature and its basis in an equitable benefit-sharing.

An entirely different option would involve a recourse by the UN General Assembly to the advisory function of the International Court of Justice, which could yield a special interpretative opinion on crucial provisions of international space law, transcending national aspirations and claims. Such an opinion could establish a new point of departure for more effective and widely accepted governance of space resource exploration and utilization.

Chapter 7
Conclusion

The global governance of space activities is inherently international. Given the status of outer space as a global commons, and in light of the explicit applicability of international law through Articles I para. 2 and III of the Outer Space Treaty (OST), it can be reasonably concluded that the rules governing celestial bodies—as well as interplanetary space—are international in nature. Accordingly, all activities in outer space, whether carried out by public or private actors, are subject to international law. The same legal framework will, of course, apply to future activities related to the use and exploitation of space mineral resources.

Commercial activities of a private nature in outer space should not be discouraged. Such activities are imposed by a new state of affairs in outer space, and it seems that they are going to expand in the near future. However, in view of the desire of some States to place the concept of 'use of space' in the context of a business perspective, the current international law of outer space may prove inadequate. Furthermore, the potential 'legal deficit' in the issue of 'space mining'—and the resulting unregulated, anarchic competition between the—technologically and economically more developed—States for sovereignty over space resources may have particularly negative implications for international security in relation to outer space, including the Moon and other celestial bodies.

The intended international regulation of space resources cannot take place without considering other aspects of space activities that *de facto* affect and are affected by the management of space resources. **Reference is made to the need to protect the space environment, particularly as it concerns planetary protection**. However, the actual extent of such protection remains to be determined—especially if an entirely anthropocentric view prevails, not only with regard to outer space and the nature of space law, but also in the face of the pressing need to accommodate powerful economic interests.

G. D. Kyriakopoulos, *Space Resource Utilization and Exploitation: The Suspended Step of International Law*, Space and Society,
https://doi.org/10.1007/978-3-031-96336-0_7

Further, the long-term sustainability of outer space activities and the "Space 2030 Agenda"—issues currently under discussion within COPUOS—are resource-related goals, which promote, by their very nature, a rational and equitable use of space resources. However, such a goal cannot be achieved through unilateral initiatives: an interesting model in this respect is provided by Article 44 of the ITU Constitution, which dictates a "rational, equitable, efficient and economic" use of the radio frequency spectrum by all radiocommunication services.

An international governance of space resource activities, with all the above characteristics, requires in the medium- or long term, in order to be effective, **the establishment of an international legal framework**. Relevant legal experience from other areas of international concern includes the governance of international airspace under the auspices of the International Civil Aviation Organization (ICAO) or the administration of Ocean seabed by the International Seabed Authority (UNCLOS, Part XI, combined with the 1994 Agreement). It is hereby proposed that the establishment of an *international* governance framework for outer space activities constitutes a *sine qua non* condition for an efficient, lawful and sustainable exploitation of outer space resources, independently of the nature of the stakeholders involved.

The path toward an international legal framework remains complex. This book has outlined: (a) the legal challenges embedded in the current international space law framework, particularly the limitations arising from the non-appropriation principle set out in Article II of the Outer Space Treaty; (b) the variety of perspectives within legal scholarship, with many interpretations reaffirming the restrictive scope of that principle; and (c) the ongoing deliberations in international fora—especially within the COPUOS Legal Subcommittee—which, while substantive, have yet to lead to a clear, constructive, and widely accepted outcome.

However, it is essential to find a way forward. If left unaddressed, the lack of clarity surrounding claims to space resources could give rise to frictions or disputes between States, with potential implications for international peace and security.

A simple agreement by the international community on **a set of "soft law" principles or guidelines**—which is currently under discussion in the Legal Subcommittee of COPUOS—will ultimately not avoid the legal deadlocks of the present, as it can only operate *intra legem*. It would, of course, represent a significant step forward, but not necessarily a comprehensive solution to the matter. Meanwhile, a process of gradually forming customary rules through existing national legislations would lead to a fragmentation of obligations, thus failing to ensure the desired universality. Therefore, as previously mentioned, the establishment of **an international conventional regime—either *ad hoc* or within the framework of a new, comprehensive treaty on outer space**—is considered the most appropriate, secure, and effective solution. The treaty-based approach could draw on existing, well-tested models, such as the international legal regime governing the exploitation of the deep seabed. However, should the international space community remain reluctant to undertake new treaty obligations, an alternative is proposed here—one which has never before seen the light of day: **A recourse to the International Court of Justice, through the mechanism of Article 96 of the UN Charter, for the issuance of an advisory opinion on the matter.**

A question to the International Court of Justice, on which it would give an advisory opinion, could help international law to take the decisive step that circumstances require in the matter under consideration. At the same time, it would make it easier for the States, for the international community as a whole, to use the resources of space not only efficiently, but also under conditions that ensure harmonious international cooperation and the maintenance of international peace and security.

Annexes

1. Treaty on Principles Governing the Activities of States in the Exploration and Use of Outer Space, including the Moon and Other Celestial Bodies
2. Agreement Governing the Activities of States on the Moon and Other Celestial Bodies
3. Proposal for the establishment of a working group for the development of an international regime for the utilization and exploitation of space resources, Working Paper by Belgium and Greece, A/AC.105/C.2/L.311, 4 March 2019
4. Questionnaire related to the discussion of item 15 on potential legal models for the exploration, exploitation, and utilization of space resources, Working Paper submitted by Greece, A/AC.105/C.2/2022/CRP.13, 28 March 2022
5. Building Blocks for the Development of an International Framework on Space Resource Activities, The Hague International Space Resources Governance Working Group, November 2019
6. The Artemis Accords—Principles for Cooperation in the Civil Exploration and Use of the Moon, Mars, Comets, and Asteroids for Peaceful Purposes

Treaty on Principles Governing the Activities of States in the Exploration and Use of Outer Space, Including the Moon and Other Celestial Bodies

The States Parties to this Treaty, Inspired by the great prospects opening up before mankind as a result of man's entry into outer space,

Recognizing the common interest of all mankind in the progress of the exploration and use of outer space for peaceful purposes,

G. D. Kyriakopoulos, *Space Resource Utilization and Exploitation: The Suspended Step of International Law*, Space and Society,
https://doi.org/10.1007/978-3-031-96336-0

Believing that the exploration and use of outer space should be carried on for the benefit of all peoples irrespective of the degree of their economic or scientific development,

Desiring to contribute to broad international cooperation in the scientific as well as the legal aspects of the exploration and use of outer space for peaceful purposes,

Believing that such cooperation will contribute to the development of mutual understanding and to the strengthening of friendly relations between States and peoples,

Recalling resolution 1962 (XVIII), entitled "Declaration of Legal Principles Governing the Activities of States in the Exploration and Use of Outer Space", which was adopted unanimously by the United Nations General Assembly on 13 December 1963,

Recalling resolution 1884 (XVIII), calling upon States to refrain from placing in orbit around the Earth any objects carrying nuclear weapons or any other kinds of weapons of mass destruction or from installing such weapons on celestial bodies, which was adopted unanimously by the United Nations General Assembly on 17 October 1963,

Taking account of United Nations General Assembly resolution 110 (II) of 3 November 1947, which condemned propaganda designed or likely to provoke or encourage any threat to the peace, breach of the peace or act of aggression, and considering that the aforementioned resolution is applicable to outer space,

Convinced that a Treaty on Principles Governing the Activities of States in the Exploration and Use of Outer Space, including the Moon and Other Celestial Bodies, will further the purposes and principles of the Charter of the United Nations,

Have agreed on the following:

Article I

The exploration and use of outer space, including the Moon and other celestial bodies, shall be carried out for the benefit and in the interests of all countries, irrespective of their degree of economic or scientific development, and shall be the province of all mankind.

Outer space, including the Moon and other celestial bodies, shall be free for exploration and use by all States without discrimination of any kind, on a basis of equality and in accordance with international law, and there shall be free access to all areas of celestial bodies.

There shall be freedom of scientific investigation in outer space, including the Moon and other celestial bodies, and States shall facilitate and encourage international cooperation in such investigation.

Article II

Outer space, including the Moon and other celestial bodies, is not subject to national appropriation by claim of sovereignty, by means of use or occupation, or by any other means.

Article III

States Parties to the Treaty shall carry on activities in the exploration and use of outer space, including the Moon and other celestial bodies, in accordance with international law, including the Charter of the United Nations, in the interest of maintaining international peace and security and promoting international cooperation and understanding.

Article IV

States Parties to the Treaty undertake not to place in orbit around the Earth any objects carrying nuclear weapons or any other kinds of weapons of mass destruction, install such weapons on celestial bodies, or station such weapons in outer space in any other manner.

The Moon and other celestial bodies shall be used by all States Parties to the Treaty exclusively for peaceful purposes. The establishment of military bases, installations and fortifications, the testing of any type of weapons and the conduct of military manoeuvres on celestial bodies shall be forbidden. The use of military personnel for scientific research or for any other peaceful purposes shall not be prohibited. The use of any equipment or facility necessary for peaceful exploration of the Moon and other celestial bodies shall also not be prohibited.

Article V

States Parties to the Treaty shall regard astronauts as envoys of mankind in outer space and shall render to them all possible assistance in the event of accident, distress, or emergency landing on the territory of another State Party or on the high seas. When astronauts make such a landing, they shall be safely and promptly returned to the State of registry of their space vehicle.

In carrying on activities in outer space and on celestial bodies, the astronauts of one State Party shall render all possible assistance to the astronauts of other States Parties.

States Parties to the Treaty shall immediately inform the other States Parties to the Treaty or the Secretary-General of the United Nations of any phenomena they discover in outer space, including the Moon and other celestial bodies, which could constitute a danger to the life or health of astronauts.

Article VI

States Parties to the Treaty shall bear international responsibility for national activities in outer space, including the Moon and other celestial bodies, whether such activities are carried on by governmental agencies or by non-governmental entities, and for assuring that national activities are carried out in conformity with the provisions set forth in the present Treaty. The activities of non-governmental entities in outer space, including the Moon and other celestial bodies, shall require authorization and continuing supervision by the appropriate State Party to the Treaty. When activities are carried on in outer space, including the Moon and other celestial bodies, by an international organization, responsibility for compliance with this Treaty shall be borne both by the international organization and by the States Parties to the Treaty participating in such organization.

Article VII

Each State Party to the Treaty that launches or procures the launching of an object into outer space, including the Moon and other celestial bodies, and each State Party from whose territory or facility an object is launched, is internationally liable for damage to another State Party to the Treaty or to its natural or juridical persons by such object or its component parts on the Earth, in air space or in outer space, including the Moon and other celestial bodies.

Article VIII

A State Party to the Treaty on whose registry an object launched into outer space is carried shall retain jurisdiction and control over such object, and over any personnel thereof, while in outer space or on a celestial body. Ownership of objects launched into outer space, including objects landed or constructed on a celestial body, and of their component parts, is not affected by their presence in outer space or on a celestial body or by their return to the Earth. Such objects or component parts found beyond the limits of the State Party to the Treaty on whose registry they are carried shall be returned to that State Party, which shall, upon request, furnish identifying data prior to their return.

Article IX

In the exploration and use of outer space, including the Moon and other celestial bodies, States Parties to the Treaty shall be guided by the principle of cooperation and mutual assistance and shall conduct all their activities in outer space, including the Moon and other celestial bodies, with due regard to the corresponding interests of all other States Parties to the Treaty. States Parties to the Treaty shall pursue studies of outer space, including the Moon and other celestial bodies, and conduct exploration of them so as to avoid their harmful contamination and also adverse changes in the environment of the Earth resulting from the introduction of extraterrestrial matter and, where necessary, shall adopt appropriate measures for this purpose. If a State Party to the Treaty has reason to believe that an activity or experiment planned by it or its nationals in outer space, including the Moon and other celestial bodies, would cause potentially harmful interference with activities of other States Parties in the peaceful exploration and use of outer space, including the Moon and other celestial bodies, it shall undertake appropriate international consultations before proceeding with any such activity or experiment. A State Party to the Treaty which has reason to believe that an activity or experiment planned by another State Party in outer space, including the Moon and other celestial bodies, would cause potentially harmful interference with activities in the peaceful exploration and use of outer space, including the Moon and other celestial bodies, may request consultation concerning the activity or experiment.

Article X

In order to promote international cooperation in the exploration and use of outer space, including the Moon and other celestial bodies, in conformity with the purposes of this Treaty, the States Parties to the Treaty shall consider on a basis of equality any requests by other States Parties to the Treaty to be afforded an opportunity to observe the flight of space objects launched by those States.

The nature of such an opportunity for observation and the conditions under which it could be afforded shall be determined by agreement between the States concerned.

Article XI

In order to promote international cooperation in the peaceful exploration and use of outer space, States Parties to the Treaty conducting activities in outer space, including the Moon and other celestial bodies, agree to inform the Secretary-General of the United Nations as well as the public and the international scientific community, to the greatest extent feasible and practicable, of the nature, conduct, locations and results of such activities. On receiving the said information, the Secretary-General of the United Nations should be prepared to disseminate it immediately and effectively.

Article XII

All stations, installations, equipment and space vehicles on the Moon and other celestial bodies shall be open to representatives of other States Parties to the Treaty on a basis of reciprocity. Such representatives shall give reasonable advance notice of a projected visit, in order that appropriate consultations may be held and that maximum precautions may be taken to assure safety and to avoid interference with normal operations in the facility to be visited.

Article XIII

The provisions of this Treaty shall apply to the activities of States Parties to the Treaty in the exploration and use of outer space, including the Moon and other celestial bodies, whether such activities are carried on by a single State Party to the Treaty or jointly with other States, including cases where they are carried on within the framework of international intergovernmental organizations.

Any practical questions arising in connection with activities carried on by international intergovernmental organizations in the exploration and use of outer space, including the Moon and other celestial bodies, shall be resolved by the States Parties to the Treaty either with the appropriate international organization or with one or more States members of that international organization, which are Parties to this Treaty.

Article XIV

1. This Treaty shall be open to all States for signature. Any State which does not sign this Treaty before its entry into force in accordance with paragraph 3 of this article may accede to it at any time.

2. This Treaty shall be subject to ratification by signatory States. Instruments of ratification and instruments of accession shall be deposited with the Governments of the Union of Soviet Socialist Republics, the United Kingdom of Great Britain and Northern Ireland and the United States of America, which are hereby designated the Depositary Governments.
3. This Treaty shall enter into force upon the deposit of instruments of ratification by five Governments including the Governments designated as Depositary Governments under this Treaty.
4. For States whose instruments of ratification or accession are deposited subsequent to the entry into force of this Treaty, it shall enter into force on the date of the deposit of their instruments of ratification or accession.
5. The Depositary Governments shall promptly inform all signatory and acceding States of the date of each signature, the date of deposit of each instrument of ratification of and accession to this Treaty, the date of its entry into force and other notices.
6. This Treaty shall be registered by the Depositary Governments pursuant to Article 102 of the Charter of the United Nations.

Article XV

Any State Party to the Treaty may propose amendments to this Treaty. Amendments shall enter into force for each State Party to the Treaty accepting the amendments upon their acceptance by a majority of the States Parties to the Treaty and thereafter for each remaining State Party to the Treaty on the date of acceptance by it.

Article XVI

Any State Party to the Treaty may give notice of its withdrawal from the Treaty 1 year after its entry into force by written notification to the Depositary Governments. Such withdrawal shall take effect 1 year from the date of receipt of this notification.

Article XVII

This Treaty, of which the Chinese, English, French, Russian and Spanish texts are equally authentic, shall be deposited in the archives of the Depositary Governments. Duly certified copies of this Treaty shall be transmitted by the Depositary Governments to the Governments of the signatory and acceding States.

IN WITNESS WHEREOF the undersigned, duly authorized, have signed this Treaty.

DONE in triplicate, at the cities of London, Moscow and Washington, D.C., the twenty-seventh day of January, one thousand nine hundred and sixty-seven.

Agreement Governing the Activities of States on the Moon and Other Celestial Bodies

The States Parties to this Agreement,

Noting the achievements of States in the exploration and use of the Moon and other celestial bodies,

Recognizing that the Moon, as a natural satellite of the Earth, has an important role to play in the exploration of outer space,

Determined to promote on the basis of equality the further development of cooperation among States in the exploration and use of the Moon and other celestial bodies,

Desiring to prevent the Moon from becoming an area of international conflict,

Bearing in mind the benefits which may be derived from the exploitation of the natural resources of the Moon and other celestial bodies,

Recalling the Treaty on Principles Governing the Activities of States in the Exploration and Use of Outer Space, including the Moon and Other Celestial Bodies, the Agreement on the Rescue of Astronauts, the Return of Astronauts and the Return of Objects Launched into Outer Space, 2 the Convention on International Liability for Damage Caused by Space Objects, 3 and the Convention on Registration of Objects Launched into Outer Space,

Taking into account the need to define and develop the provisions of these international instruments in relation to the Moon and other celestial bodies, having regard to further progress in the exploration and use of outer space,

Have agreed on the following:

Article 1

1. The provisions of this Agreement relating to the Moon shall also apply to other celestial bodies within the solar system, other than the Earth, except insofar as specific legal norms enter into force with respect to any of these celestial bodies.
2. For the purposes of this Agreement reference to the Moon shall include orbits around or other trajectories to or around it.
3. This Agreement does not apply to extraterrestrial materials which reach the surface of the Earth by natural means.

Article 2

All activities on the Moon, including its exploration and use, shall be carried out in accordance with international law, in particular the Charter of the United Nations, and taking into account the Declaration on Principles of International Law concerning Friendly Relations and Cooperation among States in accordance with the Charter of the United Nations, 6 adopted by the General Assembly on 24 October 1970, in the interest of maintaining international peace and security and promoting international cooperation and mutual understanding, and with due regard to the corresponding interests of all other States Parties.

Article 3

1. The Moon shall be used by all States Parties exclusively for peaceful purposes.
2. Any threat or use of force or any other hostile act or threat of hostile act on the Moon is prohibited. It is likewise prohibited to use the Moon in order to commit any such act or to engage in any such threat in relation to the Earth, the Moon, spacecraft, the personnel of spacecraft or manmade space objects.

3. States Parties shall not place in orbit around or other trajectory to or around the Moon objects carrying nuclear weapons or any other kinds of weapons of mass destruction or place or use such weapons on or in the Moon.
4. The establishment of military bases, installations and fortifications, the testing of any type of weapons and the conduct of military manoeuvres on the Moon shall be forbidden. The use of military personnel for scientific research or for any other peaceful purposes shall not be prohibited. The use of any equipment or facility necessary for peaceful exploration and use of the Moon shall also not be prohibited.

Article 4

1. The exploration and use of the Moon shall be the province of all mankind and shall be carried out for the benefit and in the interests of all countries, irrespective of their degree of economic or scientific development. Due regard shall be paid to the interests of present and future generations as well as to the need to promote higher standards of living and conditions of economic and social progress and development in accordance with the Charter of the United Nations.
2. States Parties shall be guided by the principle of cooperation and mutual assistance in all their activities concerning the exploration and use of the Moon. International cooperation in pursuance of this Agreement should be as wide as possible and may take place on a multilateral basis, on a bilateral basis or through international intergovernmental organizations.

Article 5

1. States Parties shall inform the Secretary-General of the United Nations as well as the public and the international scientific community, to the greatest extent feasible and practicable, of their activities concerned with the exploration and use of the Moon. Information on the time, purposes, locations, orbital parameters and duration shall be given in respect of each mission to the Moon as soon as possible after launching, while information on the results of each mission, including scientific results, shall be furnished upon completion of the mission. In the case of a mission lasting more than 60 days, information on conduct of the mission, including any scientific results, shall be given periodically, at 30-day intervals. For missions lasting more than 6 months, only significant additions to such information need be reported thereafter.
2. If a State Party becomes aware that another State Party plans to operate simultaneously in the same area of or in the same orbit around or trajectory to or around the Moon, it shall promptly inform the other State of the timing of and plans for its own operations.
3. In carrying out activities under this Agreement, States Parties shall promptly inform the Secretary-General, as well as the public and the international scientific community, of any phenomena they discover in outer

space, including the Moon, which could endanger human life or health, as well as of any indication of organic life.

Article 6

1. There shall be freedom of scientific investigation on the Moon by all States Parties without discrimination of any kind, on the basis of equality and in accordance with international law.
2. In carrying out scientific investigations and in furtherance of the provisions of this Agreement, the States Parties shall have the right to collect on and remove from the Moon samples of its mineral and other substances. Such samples shall remain at the disposal of those States Parties which caused them to be collected and may be used by them for scientific purposes. States Parties shall have regard to the desirability of making a portion of such samples available to other interested States Parties and the international scientific community for scientific investigation. States Parties may in the course of scientific investigations also use mineral and other substances of the Moon in quantities appropriate for the support of their missions.
3. States Parties agree on the desirability of exchanging scientific and other personnel on expeditions to or installations on the Moon to the greatest extent feasible and practicable.

Article 7

1. In exploring and using the Moon, States Parties shall take measures to prevent the disruption of the existing balance of its environment, whether by introducing adverse changes in that environment, by its harmful contamination through the introduction of extra-environmental matter or otherwise. States Parties shall also take measures to avoid harmfully affecting the environment of the Earth through the introduction of extraterrestrial matter or otherwise.
2. States Parties shall inform the Secretary-General of the United Nations of the measures being adopted by them in accordance with paragraph 1 of this article and shall also, to the maximum extent feasible, notify him in advance of all placements by them of radioactive materials on the Moon and of the purposes of such placements.
3. States Parties shall report to other States Parties and to the Secretary-General concerning areas of the Moon having special scientific interest in order that, without prejudice to the rights of other States Parties, consideration may be given to the designation of such areas as international scientific preserves for which special protective arrangements are to be agreed upon in consultation with the competent bodies of the United Nations.

Article 8

1. States Parties may pursue their activities in the exploration and use of the Moon anywhere on or below its surface, subject to the provisions of this Agreement.
2. For these purposes States Parties may, in particular:

   (a) Land their space objects on the Moon and launch them from the Moon;
   (b) Place their personnel, space vehicles, equipment, facilities, stations and installations anywhere on or below the surface of the Moon.

   Personnel, space vehicles, equipment, facilities, stations and installations may move or be moved freely over or below the surface of the Moon.
3. Activities of States Parties in accordance with paragraphs 1 and 2 of this article shall not interfere with the activities of other States Parties on the Moon. Where such interference may occur, the States Parties concerned shall undertake consultations in accordance with article 15, paragraphs 2 and 3, of this Agreement.

Article 9

1. States Parties may establish manned and unmanned stations on the Moon. A State Party establishing a station shall use only that area which is required for the needs of the station and shall immediately inform the Secretary-General of the United Nations of the location and purposes of that station. Subsequently, at annual intervals that State shall likewise inform the Secretary-General whether the station continues in use and whether its purposes have changed.
2. Stations shall be installed in such a manner that they do not impede the free access to all areas of the Moon of personnel, vehicles and equipment of other States Parties conducting activities on the Moon in accordance with the provisions of this Agreement or of article I of the Treaty on Principles Governing the Activities of States in the Exploration and Use of Outer Space, including the Moon and Other Celestial Bodies.

Article 10

1. States Parties shall adopt all practicable measures to safeguard the life and health of persons on the Moon. For this purpose they shall regard any person on the Moon as an astronaut within the meaning of article V of the Treaty on Principles Governing the Activities of States in the Exploration and Use of Outer Space, including the Moon and Other Celestial Bodies and as part of the personnel of a spacecraft within the meaning of the Agreement on the Rescue of Astronauts, the Return of Astronauts and the Return of Objects Launched into Outer Space.
2. States Parties shall offer shelter in their stations, installations, vehicles and other facilities to persons in distress on the Moon.

Article 11

1. The Moon and its natural resources are the common heritage of mankind, which finds its expression in the provisions of this Agreement, in particular in paragraph 5 of this article.
2. The Moon is not subject to national appropriation by any claim of sovereignty, by means of use or occupation, or by any other means.
3. Neither the surface nor the subsurface of the Moon, nor any part thereof or natural resources in place, shall become property of any State, international intergovernmental or non-governmental organization, national organization or non-governmental entity or of any natural person. The placement of personnel, space vehicles, equipment, facilities, stations and installations on or below the surface of the Moon, including structures connected with its surface or subsurface, shall not create a right of ownership over the surface or the subsurface of the Moon or any areas thereof. The foregoing provisions are without prejudice to the international regime referred to in paragraph 5 of this article.
4. States Parties have the right to exploration and use of the Moon without discrimination of any kind, on the basis of equality and in accordance with international law and the terms of this Agreement.
5. States Parties to this Agreement hereby undertake to establish an international regime, including appropriate procedures, to govern the exploitation of the natural resources of the Moon as such exploitation is about to become feasible. This provision shall be implemented in accordance with article 18 of this Agreement.
6. In order to facilitate the establishment of the international regime referred to in paragraph 5 of this article, States Parties shall inform the Secretary-General of the United Nations as well as the public and the international scientific community, to the greatest extent feasible and practicable, of any natural resources they may discover on the Moon.
7. The main purposes of the international regime to be established shall include:

   (a) The orderly and safe development of the natural resources of the Moon;
   (b) The rational management of those resources;
   (c) The expansion of opportunities in the use of those resources;
   (d) An equitable sharing by all States Parties in the benefits derived from those resources, whereby the interests and needs of the developing countries, as well as the efforts of those countries which have contributed either directly or indirectly to the exploration of the Moon, shall be given special consideration.

8. All the activities with respect to the natural resources of the Moon shall be carried out in a manner compatible with the purposes specified in paragraph 7 of this article and the provisions of article 6, paragraph 2, of this Agreement.

Article 12

1. States Parties shall retain jurisdiction and control over their personnel, vehicles, equipment, facilities, stations and installations on the Moon. The ownership of space vehicles, equipment, facilities, stations and installations shall not be affected by their presence on the Moon.
2. Vehicles, installations and equipment or their component parts found in places other than their intended location shall be dealt with in accordance with article 5 of the Agreement on the Rescue of Astronauts, the Return of Astronauts and the Return of Objects Launched into Outer Space.
3. In the event of an emergency involving a threat to human life, States Parties may use the equipment, vehicles, installations, facilities or supplies of other States Parties on the Moon. Prompt notification of such use shall be made to the Secretary-General of the United Nations or the State Party concerned.

Article 13

A State Party which learns of the crash landing, forced landing or other unintended landing on the Moon of a space object, or its component parts, that were not launched by it, shall promptly inform the launching State Party and the Secretary-General of the United Nations.

Article 14

1. States Parties to this Agreement shall bear international responsibility for national activities on the Moon, whether such activities are carried on by governmental agencies or by non-governmental entities, and for assuring that national activities are carried out in conformity with the provisions set forth in this Agreement. States Parties shall ensure that non-governmental entities under their jurisdiction shall engage in activities on the Moon only under the authority and continuing supervision of the appropriate State Party.
2. States Parties recognize that detailed arrangements concerning liability for damage caused on the Moon, in addition to the provisions of the Treaty on Principles Governing the Activities of States in the Exploration and Use of Outer Space, including the Moon and Other Celestial Bodies and the Convention on International Liability for Damage Caused by Space Objects, may become necessary as a result of more extensive activities on the Moon. Any such arrangements shall be elaborated in accordance with the procedure provided for in article 18 of this Agreement.

Article 15

1. Each State Party may assure itself that the activities of other States Parties in the exploration and use of the Moon are compatible with the provisions of this Agreement. To this end, all space vehicles, equipment, facilities, stations and installations on the Moon shall be open to other States Parties. Such States Parties shall give reasonable advance notice of a projected

visit, in order that appropriate consultations may be held and that maximum precautions may be taken to assure safety and to avoid interference with normal operations in the facility to be visited. In pursuance of this article, any State Party may act on its own behalf or with the full or partial assistance of any other State Party or through appropriate international procedures within the framework of the United Nations and in accordance with the Charter.

2. A State Party which has reason to believe that another State Party is not fulfilling the obligations incumbent upon it pursuant to this Agreement or that another State Party is interfering with the rights which the former State has under this Agreement may request consultations with that State Party. A State Party receiving such a request shall enter into such consultations without delay. Any other State Party which requests to do so shall be entitled to take part in the consultations. Each State Party participating in such consultations shall seek a mutually acceptable resolution of any controversy and shall bear in mind the rights and interests of all States Parties. The Secretary-General of the United Nations shall be informed of the results of the consultations and shall transmit the information received to all States Parties concerned.
3. If the consultations do not lead to a mutually acceptable settlement which has due regard for the rights and interests of all States Parties, the parties concerned shall take all measures to settle the dispute by other peaceful means of their choice appropriate to the circumstances and the nature of the dispute. If difficulties arise in connection with the opening of consultations or if consultations do not lead to a mutually acceptable settlement, any State Party may seek the assistance of the Secretary-General, without seeking the consent of any other State Party concerned, in order to resolve the controversy. A State Party which does not maintain diplomatic relations with another State Party concerned shall participate in such consultations, at its choice, either itself or through another State Party or the Secretary-General as intermediary.

Article 16

With the exception of articles 17–21, references in this Agreement to States shall be deemed to apply to any international intergovernmental organization which conducts space activities if the organization declares its acceptance of the rights and obligations provided for in this Agreement and if a majority of the States members of the organization are States Parties to this Agreement and to the Treaty on Principles Governing the Activities of States in the Exploration and Use of Outer Space, including the Moon and Other Celestial Bodies. States members of any such organization which are States Parties to this Agreement shall take all appropriate steps to ensure that the organization makes a declaration in accordance with the foregoing.

Article 17

Any State Party to this Agreement may propose amendments to the Agreement. Amendments shall enter into force for each State Party to the Agreement accepting the amendments upon their acceptance by a majority of the States Parties to the Agreement and thereafter for each remaining State Party to the Agreement on the date of acceptance by it.

Article 18

Ten years after the entry into force of this Agreement, the question of the review of the Agreement shall be included in the provisional agenda of the General Assembly of the United Nations in order to consider, in the light of past application of the Agreement, whether it requires revision. However, at any time after the Agreement has been in force for 5 years, the Secretary-General of the United Nations, as depositary, shall, at the request of one third of the States Parties to the Agreement and with the concurrence of the majority of the States Parties, convene a conference of the States Parties to review this Agreement. A review conference shall also consider the question of the implementation of the provisions of article 11, paragraph 5, on the basis of the principle referred to in paragraph 1 of that article and taking into account in particular any relevant technological developments.

Article 19

1. This Agreement shall be open for signature by all States at United Nations Headquarters in New York.
2. This Agreement shall be subject to ratification by signatory States. Any State which does not sign this Agreement before its entry into force in accordance with paragraph 3 of this article may accede to it at any time. Instruments of ratification or accession shall be deposited with the Secretary-General of the United Nations.
3. This Agreement shall enter into force on the 30th day following the date of deposit of the fifth instrument of ratification.
4. For each State depositing its instrument of ratification or accession after the entry into force of this Agreement, it shall enter into force on the thirtieth day following the date of deposit of any such instrument.
5. The Secretary-General shall promptly inform all signatory and acceding States of the date of each signature, the date of deposit of each instrument of ratification or accession to this Agreement, the date of its entry into force and other notices.

Article 20

Any State Party to this Agreement may give notice of its withdrawal from the Agreement 1 year after its entry into force by written notification to the Secretary-General of the United Nations. Such withdrawal shall take effect 1 year from the date of receipt of this notification.

Article 21

The original of this Agreement, of which the Arabic, Chinese, English, French, Russian and Spanish texts are equally authentic, shall be deposited with the Secretary-General of the United Nations, who shall send certified copies thereof to all signatory and acceding States.

IN WITNESS WHEREOF the undersigned, being duly authorized thereto by their respective Governments, have signed this Agreement, opened for signature at New York on the eighteenth day of December, one thousand nine hundred and seventy-nine.

United Nations A/AC.105/C.2/L.311

General Assembly

Distr.: Limited
4 March 2019

Original: English

Committee on the Peaceful Uses of Outer Space
Legal Subcommittee
Fifty-eighth session
Vienna, 1–12 April 2019

Proposal for the establishment of a working group for the development of an international regime for the utilization and exploitation of space resources

Working paper by Belgium and Greece

1. The issue of potential legal models for activities in the exploration, exploitation and utilization of space resources was discussed during the fifty-seventh session of the Legal Subcommittee of the Committee on the Peaceful Uses of Outer Space, under agenda item 15.

2. During the discussion, the delegation of Greece proposed the idea of creating an ad hoc working group of the Subcommittee with a mandate to develop and propose alternative legal solutions capable of providing the legal certainty necessary for acts of exploration, exploitation and utilization of outer space resources.[1] The proposal was supported by at least one other delegation.

3. Furthermore, under agenda item 16 (Proposals to the Committee on the Peaceful Uses of Outer Space for new items to be considered by the Legal Subcommittee at its fifty-eighth session), the delegation of Greece expressed the view that consideration of agenda item 15 should be included in the work of the Working Group on the Status and Application of the Five United Nations Treaties on Outer Space, in order to stimulate a focused debate.

4. The proposal was supported by Belgium, Brazil, China and other States. Following a proposal by Brazil, those States expressed their intention to hold consultations during the intersessional period, with a view to presenting to the Subcommittee at its fifty-eighth session, in 2019, a proposal for its consideration containing objectives and modalities for the inclusion of that item on the agenda of the Working Group on the Status and Application of the Five United Nations Treaties on Outer Space.[2]

5. The present working paper constitutes a first effort by Belgium and Greece to initiate the relevant discussion in the context of the aforementioned informal consultations.

[1] A/AC.105/1177, paras. 264 and 265.

[2] Ibid., paras. 269 and 271.

V.19-01251 (E)

6. Over the last few years, there has been intense debate on the development of activities related to the exploration, exploitation and utilization of outer space resources. This debate must take into account the existence of relevant ambitions, the attraction of significant investment and the development of the necessary technology, so that planned, public or private activities of such a character in outer space have a promising future. What is more, such activities need a solid, unequivocal legal regime, which should be formulated on the basis of the following principles of international space law in force.

I. Relevant principles of international space law in force

A. Exploration and use of outer space as the province of all humankind

7. According to the Treaty on Principles Governing the Activities of States in the Exploration and Use of Outer Space, including the Moon and Other Celestial Bodies, outer space constitutes an area beyond the jurisdiction of States. This is clear from, inter alia, article I, paragraphs 1 and 2, of the Outer Space Treaty, which state that the exploration and use of outer space shall be carried out for the benefit and in the interests of all countries and shall be the province of all humankind and that outer space shall be free for exploration and use by all States without discrimination of any kind, on a basis of equality and in accordance with international law, and there shall be free access to all areas of celestial bodies.

8. The absence of any national jurisdiction over outer space, or parts thereof, is also patent in the principle of non-appropriation, enshrined in article II of the Outer Space Treaty, pursuant to which outer space, including the Moon and other celestial bodies, is not subject to national appropriation by claim of sovereignty, by means of use or occupation or by any other means.

B. Outer space is a common space regulated by international law

9. All activities in the exploration and use of outer space are regulated by international law. This is apparent from articles I, II, III and VI of the Outer Space Treaty. Although article VI characterizes all activities of States in outer space as "national" space activities,[3] article III provides that States parties to the Treaty shall carry on activities in the exploration and use of outer space, including the Moon and other celestial bodies, in accordance with international law, including the Charter of the United Nations, in the interest of maintaining international peace and security and promoting international cooperation and understanding.

10. The applicability of international law is further reinforced by article I, in which it is stipulated that outer space, including the Moon and other celestial bodies, shall be free for exploration and use by all States without discrimination of any kind, on a basis of equality and in accordance with international law.

11. Article I of the Outer Space Treaty thus pronounces the need to adopt space law rules of an international nature to regulate the use of space (and its resources). Of course, States may authorize the space activities of their nationals. However, this personal basis for the exercise of national jurisdiction does not provide any basis for legislative jurisdiction in terms of regulating the legal status of outer space itself. It follows that the legal aspects of space resource exploitation must be regulated by international law.

[3] Article VI of the Outer Space Treaty also makes it clear that the main purpose of national space law and the obligation of supervision is for States to ensure that national activities are carried out in conformity with the provisions set forth in the Treaty.

12. With respect to the exploration and exploitation of outer space, international space law in force mainly comprises the Outer Space Treaty and the Agreement Governing the Activities of States on the Moon and Other Celestial Bodies. To date, 18 States have ratified the Moon Agreement, which provides, in its article 11, for the obligation among States parties to undertake to establish an international regime of exploitation of the natural resources of the Moon, "as such exploitation is about to become feasible". For the majority of the members of the Committee that have not yet ratified the Agreement, the international norms applicable to the exploration and exploitation of outer space are the relevant provisions of the Outer Space Treaty. In this context, particular attention must be given to articles I and II of the Outer Space Treaty.

C. Global governance of outer space activities is international in character

13. In the context of the present working paper, the term "governance" is defined as the way that outer space is managed at the highest level and the systems for doing so. Given the nature of outer space as a global commons and taking into consideration the role of international law in this respect,[4] it can be validly concluded that the rules in force with respect to the governance of celestial bodies and of the void of outer space (including orbits around the Earth or other celestial bodies) are part of international law.

14. It follows that activities in outer space, whether public or private in nature, are governed by international law.

15. Such international governance, in order to be effective, requires, in the medium or long term, the establishment of an international institutional framework. Past legal experience in this respect includes the administration of international airspace by the International Civil Aviation Organization (through the recognition, by its member States, of a series of functional jurisdictions inside the so-called "flight information regions"), the administration of the seabed by the International Seabed Authority (United Nations Convention on the Law of the Sea, part XI, in combination with the 1994 Agreement relating to the Implementation of Part XI), the frequency spectrum management regime of the International Telecommunication Union and the legal regime governing Antarctica (under the 1959 Antarctic Treaty).

16. The need for an international legal regime for space resource exploitation also arises from the fact that national approaches to space resource exploitation are bound to result in conflicts between competing players, if left to evolve on their own without international guidance. Hence, even if there is no legal objection to States interpreting at will their international obligations under the Outer Space Treaty when regulating space resources, there is still a clear need for an international institutional framework to regulate competing activities. In order for such a framework to be effective, it would have to be focused on the main purposes described in article 11, paragraph 7, of the Moon Agreement, the value of which is greater and goes beyond any views on the ratification of the Moon Agreement. Those purposes include the following: (a) the orderly and safe development of natural resources from outer space; (b) the rational management of those resources; (c) the expansion of opportunities in the use of those resources; and (d) an equitable sharing by all States in the benefits derived from those resources, whereby the interests and needs of the developing countries, as well as the efforts of those countries which have contributed either directly or indirectly to the exploration of outer space, shall be given special consideration.

17. It is therefore reasonable to conclude that the establishment of an international governance framework for outer space activities constitutes a sine qua non for the efficient, lawful and sustainable exploitation of outer space resources that is independent of the nature of the stakeholders involved.

[4] A/AC.105/1177, para. 264.

D. The outer space treaties impose enhanced international cooperation with respect to the exploration and use of outer space

18. An international governance scheme for outer space activities (and, consequently, for any attempt to exploit resources in outer space) presupposes the establishment of enhanced international cooperation. International cooperation is at the heart and in the spirit of the Outer Space Treaty; articles I, III, V and IX–XII are based on or refer to it. What is more, the importance of international cooperation in outer space is reflected, constantly and continuously, in the animus of the international space community, as it is often repeated in numerous General Assembly resolutions. For instance, in the Declaration on the fiftieth anniversary of the Treaty on Principles Governing the Activities of States in the Exploration and Use of Outer Space, including the Moon and Other Celestial Bodies, adopted by the General Assembly in its resolution 72/78, Member States emphasized the constantly evolving and increasingly multifaceted nature of international cooperation in the peaceful uses of outer space, with fundamentally complex scientific and technological advancements in the space field and an increasing variety of actors in the space arena, and therefore encouraged the building of stronger partnerships, cooperation and coordination. Also in that Declaration, Member States recognized the need for further promotion of international cooperation in the peaceful uses of outer space, taking into account the particular needs of developing countries.

19. Furthermore, in its resolution 72/77, entitled "International cooperation in the peaceful uses of outer space", the General Assembly reaffirmed the importance of international cooperation in developing the rule of international law, including the relevant norms of international space law and their important role in international cooperation for the exploration and use of outer space for peaceful purposes, and of the widest possible adherence to international treaties promoting the peaceful uses of outer space in order to meet emerging new challenges, especially for developing countries.[5]

20. The requirement for international cooperation in the specific field of space resource exploitation calls for increased multilateralism in planning, organizing, extracting and exploiting space resources; article 11 of the Moon Agreement, even though currently only ratified by 18 States, can provide a useful starting point for further discussions concerning the establishment of an international regime for the exploitation of space resources in accordance with established principles of international space law.

II. The way forward

21. Constructive discussions aimed at the solution of complex legal problems require both a shared factual basis and a well-structured workplan implemented through a broadly recognized, inclusive platform with the appropriate legal mandate.

22. It is proposed that the efficient and orderly management of space resource exploitation should proceed on the basis of a structured, multi-year and multilayered discussion concerning its legal regulation at the international level. This discussion should be based on a collection of up-to-date information from all relevant stakeholders in this emerging field, as the first step towards arriving at a workable

[5] See also the following General Assembly resolutions: 51/122 of 13 December 1996, 54/68 of 6 December 1999, 59/2 of 20 October 2004, 61/110 and 61/111 of 14 December 2006, 62/101 of 17 December 2007, 62/217 of 22 December 2007, 65/97 of 10 December 2010, 65/271 of 7 April 2011, 66/71 of 9 December 2011, 67/113 of 18 December 2012, 68/50 of 5 December 2013, 68/74 and 68/75 of 11 December 2013, 69/85 of 5 December 2014, 70/1 of 25 September 2015, 70/82 of 9 December 2015, 70/230 of 23 December 2015 and 71/90 of 6 December 2016.

regime that facilitates the exploitation of space resources in line with existing international space law.

23. Considering the current state of play of the space resources industry and the fragmented nature of technical, economic and scientific information concerning space resources, it is therefore deemed indispensable that legal discussions start with the organization of one or more information-gathering rounds to ensure a common understanding and shared factual basis.

24. The interpretation of international legal principles, such as those contained in the Outer Space Treaty, remains the prerogative of States. For this reason, and taking into account the mandate of the Committee on the Peaceful Uses of Outer Space, specified in General Assembly resolution 1472 (XIV) of 12 December 1959, to study the nature of legal problems that may arise from the exploration of outer space, and in that connection, also taking into account the Committee's role in both developing legal instruments, be they binding or non-binding, and providing a platform for constructive discussions on these issues, Belgium and Greece consider the Committee to be the primary forum for intergovernmental discussions and negotiations on the future regime for space resource exploitation. This is further supported by the observation that a major source of legal uncertainty concerning the regulation of this activity stems from the fact that States have different views on how to interpret the principles laid down in legal instruments that were developed by the States members of this same Committee, which is therefore in a unique position to offer pertinent clarifications.

25. In the light of the above, it appears that the so-called "new space" activities in outer space, in particular the emergence of a strong tendency for private space resource exploitation, require a significant effort to re-evaluate the collective body of space law, as incorporated in the five United Nations treaties on outer space.

26. Given the permanent and unequivocal affirmation by the international community of the fundamental principles enshrined in the Outer Space Treaty, such a re-evaluation would not be aimed at challenging the validity of these principles, but merely at shedding more light on their scope of application. This discussion obviously enters into the realm of the "status and application" of the United Nations treaties on outer space and, consequently, inserting the "potential models" issue into the work of the Working Group on the Status and Application of the Five United Nations Treaties on Outer Space would certainly not only have a positive impact on the search for a viable international legal regime for space resource exploitation that is in conformity with international law and for the benefit and in the interests of all countries, irrespective of their degree of economic or scientific development, but would also lead to a better understanding of the legal principles that govern the space adventure of humankind.

27. However, potential concerns in relation to the already sizeable work of the Working Group on the Status and Application of the Five United Nations Treaties on Outer Space and its capability to handle one more important issue must be taken into consideration, particularly in the light of the disparate legal positions expressed during the discussions in 2017 and 2018 under the agenda item on general exchange of views on potential legal models for activities in the exploration, exploitation and utilization of space resources, which have demonstrated the need for a constructive, multi-year process on space resource regulation within the Committee.

28. It follows that the establishment of an ad hoc working group on this matter should be provided as an alternative and preferable option. Indeed, the importance of arriving at a sustainable and efficient framework of space resource regulations and the contentious nature of the issues at stake are best served by the establishment of such a separate working group under the current item on space resources, rather than merging these complex issues with discussions of the Working Group on the Status and Application of the Five United Nations Treaties on Outer Space.

29. In accordance with established practice within the Legal Subcommittee, such a future working group on space resources would need to have a well-defined scope of work and proceed on the basis of a commonly agreed timeline to produce a report addressing specific issues identified in advance by the States members of the Committee. Some States members of the Committee have already put forward documents identifying such specific issues (for example, the working paper prepared by Belgium (A/AC.105/C.2/2018/CRP.8)), and those issues form the basis of the following preliminary list of issues that can be considered as deserving priority treatment by any future working group on space resources:

(a) Definition of terms and application of general principles on the exploration and use of outer space to the developing activity of space resource exploitation;

(b) Relationship with other international legal regimes on the exploitation of natural resources in international areas, such as the orbit and frequency regime of the International Telecommunication Union and the deep seabed regime;

(c) Identification and determination of the legal relevance of emerging State and other practice concerning space resources;

(d) Exclusive nature of future rights concerning space resources;

(e) Temporal and geographical delimitation of claims over areas containing space resources to support an efficient and rational use of those resources;

(f) Scientific collection of data and information-sharing obligations concerning activities of space resource exploitation;

(g) Benefit-sharing of space resource activities, taking into account the contributions of resource-collecting States and the needs of developing and non-spacefaring countries;

(h) Coordination of competing claims and dispute settlement;

(i) Institutional framework for space resource management;

(j) Appropriate means to ensure the sustainability and environmental compatibility of space resource exploitation activities.

30. As noted, the discussion of these and other issues should be informed by the views of different expert groups comprising scientific, economic, technical and legal experts in order to establish a shared factual basis from which to proceed. In this respect, much can be learned from the experience of the Working Group on the Long-term Sustainability of Outer Space Activities, which relied in large part on expert groups comprising scientific, technical, policy and legal experts nominated by member States and international intergovernmental organizations with permanent observer status in the Committee. In contrast to the diplomatic negotiation format of the Committee discussions, the expert groups were deliberative forums in which the technical experts from a wide range of countries exchanged their views and experience and proposed draft guidelines for the consideration of the Working Group.

31. The novel nature of the activities under discussion and their expected impact also requires that any future working group on space resources coordinate with other actors. In particular, the potentially disruptive economic impact of space resource exploitation activities on existing global inequalities indicates a pressing need to cooperate with existing mechanisms working on the implementation and realization of the Sustainable Development Goals.

32. In this regard, it is important to ensure appropriate coordination between the proposed working group on space resources and the working group established under the new item on the Committee's agenda, entitled "'Space2030' agenda", which is to remain on the Committee's agenda until its sixty-third session, in 2020 (A/73/20, paras. 359 and 360).

33. Finally, a comprehensive workplan for the regulation of space resource exploitation should also aim to incorporate the legal positions of those States that

have ratified the Moon Agreement. While the possibility must be recognized that the legal situation of those States may differ from that of other States that have not ratified the Moon Agreement, current developments may be interpreted as meeting the condition for the activation of the obligations of States parties to the Agreement to undertake to establish an international regime on space resource exploitation, as such exploitation is about to become feasible (Moon Agreement, art. 11, para. 5). To avoid the establishment of separate international regimes, any future working group on space resources must incorporate institutional mechanisms to properly align its discussions with possible parallel discussions among States parties to the Moon Agreement in this respect.

A/AC.105/C.2/2022/CRP.13*

28 March 2022

English only

Committee on the Peaceful Uses of Outer Space
Legal Subcommittee
Sixty-first session
Vienna, 28 March–8 April 2022
Item 15 of the provisional agenda**
General exchange of views on potential legal models for activities in the exploration, exploitation and utilization of space resources

Questionnaire related to the discussion of item 15 on potential legal models for the exploration, exploitation, and utilization of space resources

Working Paper submitted by Greece

I. Introduction and Background

1. At its 60th Session in 2021, the Legal Subcommittee of the UN COPUOS, following a series of working papers that have been submitted to the LSC since 2019 and advanced the relevant debate, decided, on the basis of the reports provided by the Moderator and Vice-Moderator of the scheduled informal consultations and on the progress made in those consultations, to establish, under a five-year workplan, a working group under the agenda item on the general exchange of views on potential legal models for activities in exploration, exploitation and utilization of space resources, with Mr. Misztal as Chair and Mr. Freeland as Vice-Chair of the working group.[1]

2. In order to facilitate the exchange of views among Member States on the subject matter in the context of the working group, Greece submits to its Chair and Vice-Chair a questionnaire for States members of the Committee, considering that this may broaden the relevant discussion and contribute positively to the creation of constructive convergences within the working group. The proposed questionnaire might equally be a component of the co-Chairs' five year Work Plan.

* Reissued for technical reasons on 29 March 2022.

** A/AC.105/C.2/L.319.

[1] See Report of the Legal Subcommittee on its sixtieth session, held in Vienna from 31 May to 11 June 2021, A/AC.105/1243, para. 255.

V.22-01720 (E)

II. Questionnaire related to the discussion on potential legal models for the exploration, exploitation and utilization of space resources

1. Is the existing international legal framework for outer space activities (mainly the five United Nations Treaties) sufficient to regulate the exploration, exploitation and utilization of space resources (a) or is an *ad hoc* framework required (b)?

1.1 If the answer to (a) is in the affirmative, which existing principles and provisions provide for an adequate regulation of the matter?

1.2 If the answer to (b) is in the affirmative, what should be the main characteristics of *an ad hoc* framework?

1.3 Further, this framework:

(a) Should it have the characteristics of a legal regime? If so, should the regime in question be regulated by international law or at the level of domestic law?

(b) Should it have the characteristics of an ex aequo et bono regime within the meaning of Article 38 para. 2 of the Statute of the International Court of Justice? If this should be the case, on which principles should such a regime be based?

2. Can the issue of exploration, exploitation and utilization of space resources be considered to fall within the scope of Article II of the Outer Space Treaty (non-appropriation principle)? In such a case, and bearing in mind that a framework for the regulation of resources is being explored.

2.1 Should Article II of the Outer Space Treaty be subject to a commonly accepted interpretation?

2.2 Should Article II of the Outer Space Treaty be amended?

2.3 Should there be a review of the concept of celestial bodies, so as to narrow the scope of Article II (e.g. as regards comets and asteroids)?

3. Should the issue of exploration, exploitation and utilization of space resources be considered with a view to an all-inclusive regulation, or should there be different approaches on a case-by-case basis (for example, in relation to orbits around the earth, to *in situ* resource utilization or to space mining for commercial exploitation)?

4. Should a legal framework for the exploration, exploitation and utilization of space resources promote an equitable access to said resources, for all States?

5. Could the Moon Agreement provide a suitable basis for the development of a legal regime for the exploration, exploitation and utilization of space resources?

6. In the context of the debate on a framework for the exploration, exploitation and utilization of space resources, should the interaction of this issue with the long-term sustainability of outer space activities be considered? In general, should the issue be considered in the light of environmental considerations, in accordance with the letter and spirit of Article IX of the Outer Space Treaty?

7. Could the existing international legal regime for mineral exploitation in the Area (international seabed) possibly serve as a model for the exploration, exploitation and utilization of space resources?

8. Are the technical and scientific aspects of the exploration, exploitation and utilization of space resources considered as factors that affect the development of a framework? If the answer is in the affirmative, what is the best way to make the relevant feedback available to the working group on space resources?

BUILDING BLOCKS FOR THE DEVELOPMENT OF AN INTERNATIONAL FRAMEWORK ON SPACE RESOURCE ACTIVITIES

Introduction

The utilization of space resources has great potential for the future of humankind. In order to create an enabling environment for space resource activities, The Hague International Space Resources Governance Working Group was created to promote international cooperation and multi-stakeholder dialogue. It has designed the building blocks below to lay the groundwork for international discussions on the potential development of an international framework, without prejudice to its form and structure. The Working Group adopted the Building Blocks on 12 November 2019. Pending the adoption and the operationalization of the international framework, States, international organizations, and non-governmental entities are encouraged to consider and use the Building Blocks.

Guided by the principle of adaptive governance, the Working Group considered it neither necessary nor feasible to attempt to comprehensively address space resource activities in the building blocks: space resource activities should be incrementally addressed at the appropriate time on the basis of contemporary technology and practices.

A commentary on the development of the Building Blocks with information on each of the Building Blocks will be made available at the beginning of 2020. Information about its publication will be posted on the website of the Working Group.[1]

The Working Group hopes that its activities will complement other efforts at the national, regional and global level to address space resource activities.

1. Objective

1.1 The international framework should create an enabling environment for space resource activities that takes into account all interests and benefits all countries and humankind.

1.2 To achieve this objective, the international framework should:

a) Identify and define the relationship of space resource activities with existing international space law, including the provisions of the United Nations treaties on outer space;

b) Propose recommendations for the consideration of States for the application or development of domestic frameworks;

c) Propose recommendations for the consideration of international organizations for the application or development of internal frameworks;

d) Promote the identification of best practices by States, international organizations and non-governmental entities.

2. Definition of key terms

2.1 Space resource: an extractable and/or recoverable abiotic resource *in situ* in outer space;[2]

[1] https://www.universiteitleiden.nl/en/law/institute-of-public-law/institute-of-air-space-law/the-hague-space-resources-governance-working-group

[2] According to the understanding of the Working Group, this includes mineral and volatile materials, including water, but excludes (a) satellite orbits; (b) radio spectrum; and (c) energy from the sun except when collected from unique and scarce locations.

2.2 Utilization of space resources: the recovery of space resources and the extraction of raw mineral or volatile materials therefrom;[3]
2.3 Space resource activity: an activity conducted in outer space for the purpose of searching for space resources, the recovery of those resources and the extraction of raw mineral or volatile materials therefrom, including the construction and operation of associated extraction, recovery, processing and transportation systems;
2.4 Space object: an object launched into outer space from Earth, including component parts thereof as well as its launch vehicle and parts thereof;
2.5 Space-made product: a product made in outer space wholly or partially from space resources;[4]
2.6 Operator: a governmental, international or non-governmental entity conducting space resource activities.

3. Scope

3.1 The international framework should address States and international organizations, and could provide for the regulation of the conduct of States, international organizations, and non-governmental entities.
3.2 The international framework should address space resource activities within the solar system.

4. Principles

4.1 The international framework should be consistent with international law.
4.2 The international framework should be designed to:
- a) Adhere to the principle of adaptive governance by incrementally regulating space resource activities at the appropriate time;
- b) Promote compatibility and predictability of domestic frameworks of States and internal frameworks of international organizations;
- c) Contribute to sustainable development;
- d) Prevent disputes arising out of space resource activities;
- e) Promote and secure the orderly and safe utilization of space resources;
- f) Promote the sustainable, rational, efficient and economic use of space resources;
- g) Promote the use of sustainable technology;
- h) Provide legal certainty and predictability for operators;
- i) Take into particular account the needs of developing countries;
- j) Take into particular account the needs of science;
- k) Take into particular account the contributions of pioneer operators.

4.3 The international framework should provide that:
- a) Space resources shall be used exclusively for peaceful purposes;
- b) Space resource activities shall be carried out for the benefit and in the interests of all countries and humankind irrespective of their degree of economic and scientific development;
- c) Appropriate international consultations shall be undertaken in accordance with Article IX OST[5] if there is a reason to believe that any potentially harmful interference may be caused;

[3] According to the understanding of the Working Group, this excludes secondary utilization of space resources, i.e. (a) utilization of raw materials derived from space resources; and (b) marketing and distribution of space resources.

[4] According to the understanding of the Working Group, this excludes raw mineral and volatile materials, including water, irrespective of form.

[5] 1967 Treaty on Principles Governing the Activities of States in the Exploration and Use of Outer Space, including the Moon and Other Celestial Bodies.

d) International cooperation in space resource activities shall be conducted in accordance with international law.

5. International responsibility for space resource activities
The international framework should provide that:

a) States shall bear international responsibility for national space resource activities, whether such activities are carried out by governmental agencies or non-governmental entities, and for ensuring that such activities are carried out in conformity with the international framework;
b) Non-governmental space resource activities shall require prior authorization and continuing supervision by the appropriate State;
c) When space resource activities are carried out by an international organization, responsibility for compliance with the international framework shall be borne by the international organization and by the States participating in such organization.

6. Jurisdiction and control over space-made products used in space resource activities
The international framework should provide that States have jurisdiction and control over any space-made products used in the space resource activities for which they are responsible.

7. Priority rights
The international framework should enable the attribution of priority rights to an operator to search for and/or recover space resources for a maximum period of time and a maximum area upon registration in an international registry, and provide for the international recognition of such priority rights. The attribution, duration, and the area of the priority right should be determined on the basis of the specific circumstances of a proposed space resource activity.

8. Resource rights

8.1 The international framework should ensure that resource rights over raw mineral and volatile materials extracted from space resources, as well as products derived therefrom, can lawfully be acquired through domestic legislation, bilateral agreements and/or multilateral agreements.
8.2 The international framework should enable the mutual recognition between States of such resource rights.
8.3 The international framework should ensure that the utilization of space resources is carried out in accordance with the principle of non-appropriation under Article II OST.

9. Due regard for corresponding interests of all countries and humankind
The international framework should provide that States and international organizations responsible for space resource activities shall give due regard to the corresponding interests of all countries and humankind.

10. Avoidance and mitigation of potentially harmful impacts resulting from space resource activities
Taking into account the current state of technology, the international framework should provide that States and international organizations responsible for space resource

activities shall adopt appropriate measures with the aim of avoiding and mitigating potentially harmful impacts, including:

a) Risks to the safety of persons, the environment or property;
b) Damage to persons, the environment or property;
c) Adverse changes in the environment of the Earth, taking into account internationally agreed planetary protection policies;
d) Harmful contamination of celestial bodies, taking into account internationally agreed planetary protection policies;
e) Harmful contamination of outer space;
f) Harmful effects of the creation of space debris;
g) Harmful interference with other on-going space activities, including other space resource activities;
h) Changes to designated and internationally endorsed outer space natural or cultural heritage sites;
i) Adverse changes to designated and internationally endorsed outer space sites of scientific interest.

11. Technical standards for, prior review of, and safety zones around space resource activities

11.1 The international framework should provide that States and international organizations shall require the conduct of a review prior to a decision to proceed with a space resource activity to ascertain that such an activity is carried out in a safe manner to avoid harmful impacts.

11.2 The international framework should encourage the development of:

a) Procedures to ensure that equipment, operational procedures, and processes applied in space resource activities avoid harmful impacts;
b) Methodologies to assess that equipment, operational procedures, and processes applied in space resource activities meet common technical standards (conformity assessment);
c) Technical standards for equipment, operational procedures, and processes applied in space resource activities (standardization).

11.3 Taking into account the principle of non-appropriation under Article II OST, the international framework should permit States and international organizations responsible for space resource activities to establish a safety zone, or other area-based safety measure, around an area identified for a space resource activity as necessary to assure safety and to avoid any harmful interference with that space resource activity. Such safety measure shall not impede the free access, in accordance with international law, to any area of outer space by personnel, vehicles and equipment of another operator. In accordance with the area-based safety measure, a State or international organization may restrict access for a limited period of time, provided that timely public notice has been given setting out the reasons for such restriction.

11.4 The international framework should provide that appropriate international consultations are undertaken in case of possible overlap of safety zones or conflicts involving the freedom of access recognized by international law.

12. Monitoring and redressing harmful impacts resulting from space resource activities

12.1 The international framework should provide that States and international organizations shall ensure monitoring of any harmful impacts resulting from space resource activities for which they are responsible.

12.2 If a harmful impact resulting from a space resource activity occurs, or is reasonably expected to occur, the international framework should provide that the State or international organization responsible for the space resource activity shall implement measures to respond to such harmful impact (response measures) and consider whether the space resource activity should be adjusted or terminated (adaptive management).

13. Sharing of benefits arising out of the utilization of space resources

13.1 Bearing in mind that the exploration and use of outer space shall be carried out for the benefit and in the interests of all countries and humankind, the international framework should provide that States and international organizations responsible for space resource activities shall provide for benefit-sharing through the promotion of the participation in space resource activities by all countries, in particular developing countries. Benefits may include, but not be limited to, enabling, facilitating, promoting, and fostering:

a) The development of space science and technology and of its applications;
b) The development of relevant and appropriate capabilities in interested States;
c) Cooperation and contribution in education and training;
d) Access to and exchange of information;
e) Incentivization of joint ventures;
f) The exchange of expertise and technology among States on a mutually acceptable basis;
g) The establishment of an international fund.

13.2 The international framework should not require compulsory monetary benefit-sharing.

13.3 Operators should be encouraged to provide for benefit-sharing.

14. Registration and sharing of information

The international framework should provide that States and international organizations shall:

a) Register priority rights of an operator to search and/or recover space resources in accordance with the international framework;
b) Give advance notification of space resource activities, including any area-based safety measure associated with them, for which they are responsible through an international database;
c) Register space objects in accordance with the REG,[6] United Nations General Assembly Resolution 1721 B (XVI),[7] or Article XI OST, taking into account United Nations General Assembly Resolution 62/101;[8]
d) Notify frequency assignments for recording in the Master International Frequency Register in accordance with the Radio Regulations of the International Telecommunication Union;
e) Provide, taking into account Article XI OST and the legitimate interests of operators, information and best practices on the prior authorization and continuing supervision of space resource activities for which they are responsible through an international database, including:

[6] 1975 Convention on the Registration of Objects Launched into Outer Space.

[7] 1961 Resolution 1721 B (XVI) adopted by the United Nations General Assembly, International Cooperation on the Peaceful Uses of Outer Space.

[8] 2008 Resolution 62/101 adopted by the United Nations General Assembly, Recommendations on enhancing the practice of States and international intergovernmental organizations in registering space objects.

i. The purposes, locations, orbital parameters, and duration of space resource activities;
ii. The nature, conduct, and locations of space resource activities and associated logistic activities, for example deployment of stations, installations, equipment and vehicles;
iii. The results of space resource activities;
iv. Any phenomena discovered in outer space which could endanger terrestrial life or health, as well as of any indication of extraterrestrial life;
v. Any harmful impacts resulting from space resource activities for which they are responsible and the measures planned or implemented to redress such impacts;

f) Notify the termination of space resource activities for which they are responsible through an international database together with a statement on the condition of the area where the space resource activity was carried out, including the presence of any space objects or space-made products, or parts thereof.

15. Provision of assistance in case of distress

The international framework should provide for the applicability of Article V OST and the ARRA[9] to persons involved in space resource activities.

16. Liability in case of damage resulting from space resource activities

16.1 The international framework should provide for the applicability of Articles VI and VII OST and the LIAB[10] to damage resulting from space resource activities.

16.2 The international framework should encourage initiatives of operators to provide, individually or collectively, compensation for damage resulting from their space resource activities.

17. Visits relating to space resource activities

The international framework should provide for the applicability of Article XII OST, taking into account the legitimate interests of operators, including safety of operations and protection of intellectual property.

18. Institutional arrangements

The international framework should provide for:

a) The establishment and maintenance of a publicly available international registry for registering priority rights of an operator to search and/or recover space resources;
b) The establishment and maintenance of an international database, in addition to the international registry, for making publicly available:
   i. Advance notifications of space resource activities, including any area-based safety measures;
   ii. Information and best practices;
   iii. The list of designated and internationally endorsed outer space natural and cultural heritage sites; and
   iv. The list of designated and internationally endorsed sites of scientific interest;

[9] 1968 Agreement on the Rescue of Astronauts, the Return of Astronauts and the Return of Objects Launched into Space.

[10] 1972 Convention on International Liability for Damage Caused by Space Objects.

   v. Information and best practices on the prior authorization and continuing supervision of space resource activities for which States and international organizations are responsible;
   vi. Notifications of the termination of space resource activities for which States and international organizations are responsible.

c) The designation or establishment of an international body or bodies responsible for:
   i. The consideration and promotion of best practices;
   ii. The listing of designated and internationally endorsed outer space natural and cultural heritage sites, and sites of scientific interest;
   iii. The monitoring and review of the implementation of the international framework; and
   iv. The governance of the international registry, the international database and any other mechanism that may be established for the implementation of the international framework.

19. Settlement of disputes
The international framework should encourage recourse by States, international organizations and operators to the resolution of disputes through adjudicatory, non-adjudicatory or hybrid mechanisms, for example by developing procedures for consultation or promoting the use of the 2011 Permanent Court of Arbitration Optional Rules for Arbitration of Disputes Relating to Outer Space Activities.

20. Monitoring and review
Mechanisms should be developed for monitoring implementation of the international framework, for example on the basis of reports of States and international organizations, as well as for its review and further development consistent with the principle of adaptive governance.

The Artemis Accords

Principles for Cooperation in the Civil Exploration and Use of the Moon, Mars, Comets, and Asteroids for Peaceful Purposes

Table of Contents

The Signatories to these Accords;

RECOGNIZING their mutual interest in the exploration and use of outer space for peaceful purposes, and **UNDERSCORING** the continuing importance of existing bilateral space cooperation agreements;

NOTING the benefit for all humankind to be gained from cooperating in the peaceful use of outer space;

USHERING in a new era of exploration, more than 50 years after the historic Apollo 11 Moon landing and more than 20 years after the establishment of a continuous human presence aboard the International Space Station;

SHARING a common spirit and the ambition that the next steps of humanity's journey in space inspire current and future generations to explore the Moon, Mars, and beyond;

BUILDING upon the legacy of the Apollo program, which benefited all of humankind, the Artemis program will land the first woman and next man on the surface of the Moon and establish, together with international and commercial partners, the sustainable human exploration of the solar system;

CONSIDERING the necessity of greater coordination and cooperation between and among established and emerging actors in space;

RECOGNIZING the global benefits of space exploration and commerce;

ACKNOWLEDGING a collective interest in preserving outer space heritage;

AFFIRMING the importance of compliance with the *Treaty on Principles Governing the Activities of States in the Exploration and Use of Outer Space, Including the Moon and Other Celestial Bodies*, opened for signature on January 27, 1967 ("Outer Space Treaty") as well as the *Agreement on the Rescue of Astronauts, the Return of Astronauts and the Return of Objects Launched into Outer Space*, opened for signature on April 22, 1968 ("Rescue and Return Agreement"), the *Convention on International Liability for Damage Caused by Space Objects*, opened for signature on March 29, 1972 ("Liability Convention"), and the *Convention on Registration of Objects Launched into Outer Space,* opened for signature on January 14, 1975 ("Registration Convention"); as well as the benefits of coordination via multilateral forums, such as the United Nations Committee on the Peaceful Uses of Outer Space ("COPUOS"), to further efforts toward a global consensus on critical issues regarding space exploration and use; and

DESIRING to implement the provisions of the Outer Space Treaty and other relevant international instruments and thereby establish a political understanding regarding mutually beneficial practices for the future exploration and use of outer space, with a focus on activities conducted in support of the Artemis Program;

COMMIT to the following principles:

Section 1: Purpose and Scope

The purpose of these Accords is to establish a common vision via a practical set of principles, guidelines, and best practices to enhance the governance of the civil exploration and use of outer space with the intention of advancing the Artemis

Program. Adherence to a practical set of principles, guidelines, and best practices in carrying out activities in outer space is intended to increase the safety of operations, reduce uncertainty, and promote the sustainable and beneficial use of space for all humankind. The Accords represent a political commitment to the principles described herein, many of which provide for operational implementation of important obligations contained in the Outer Space Treaty and other instruments.

The principles set out in these Accords are intended to apply to civil space activities conducted by the civil space agencies of each Signatory. These activities may take place on the Moon, Mars, comets, and asteroids, including their surfaces and subsurfaces, as well as in orbit of the Moon or Mars, in the Lagrangian points for the Earth-Moon system, and in transit between these celestial bodies and locations. The Signatories intend to implement the principles set out in these Accords through their own activities by taking, as appropriate, measures such as mission planning and contractual mechanisms with entities acting on their behalf.

Section 2: Implementation

1. Cooperative activities regarding the exploration and use of outer space may be implemented through appropriate instruments, such as Memoranda of Understanding, Implementing Arrangements under existing Government-to-Government Agreements, Agency-to-Agency arrangements, or other instruments. These instruments should reference these Accords and include appropriate provisions for implementing the principles contained in these Accords.
   (a) In the instruments described in this Section, the Signatories or their subordinate agencies should describe the nature, scope, and objectives of the civil cooperative activity;
   (b) The Signatories' bilateral instruments referred to above are expected to contain other provisions necessary to conduct such cooperation, including those related to liability, intellectual property, and the transfer of goods and technical data;
   (c) All cooperative activities should be carried out in accordance with the legal obligations applicable to each Signatory; and
   (d) Each Signatory commits to taking appropriate steps to ensure that entities acting on its behalf comply with the principles of these Accords.

Section 3: Peaceful Purposes

The Signatories affirm that cooperative activities under these Accords should be exclusively for peaceful purposes and in accordance with relevant international law.

Section 4: Transparency

The Signatories are committed to transparency in the broad dissemination of information regarding their national space policies and space exploration plans in accordance with their national rules and regulations.

The Signatories plan to share scientific information resulting from their activities pursuant to these Accords with the public and the international scientific community on a good-faith basis, and consistent with Article XI of the Outer Space Treaty.

Section 5: Interoperability

The Signatories recognize that the development of interoperable and common exploration infrastructure and standards, including but not limited to fuel storage and delivery systems, landing structures, communications systems, and power systems, will enhance space-based exploration, scientific discovery, and commercial utilization. The Signatories commit to use reasonable efforts to utilize current interoperability standards for space-based infrastructure, to establish such standards when current standards do not exist or are inadequate, and to follow such standards.

Section 6: Emergency Assistance

The Signatories commit to taking all reasonable efforts to render necessary assistance to personnel in outer space who are in distress, and acknowledge their obligations under the Rescue and Return Agreement.

Section 7: Registration of Space Objects

For cooperative activities under these Accords, the Signatories commit to determine which of them should register any relevant space object in accordance with the Registration Convention. For activities involving a non-Party to the Registration Convention, the Signatories intend to cooperate to consult with that non-Party to determine the appropriate means of registration.

Section 8: Release of Scientific Data

1. The Signatories retain the right to communicate and release information to the public regarding their own activities. The Signatories intend to coordinate with each other in advance regarding the public release of information that relates to the other Signatories' activities under these Accords in order to provide appropriate protection for any proprietary and/or export-controlled information.

2. The Signatories are committed to the open sharing of scientific data. The Signatories plan to make the scientific results obtained from cooperative activities under these Accords available to the public and the international scientific community, as appropriate, in a timely manner.
3. The commitment to openly share scientific data is not intended to apply to private sector operations unless such operations are being conducted on behalf of a Signatory to the Accords.

Section 9: Preserving Outer Space Heritage

1. The Signatories intend to preserve outer space heritage, which they consider to comprise historically significant human or robotic landing sites, artifacts, spacecraft, and other evidence of activity on celestial bodies in accordance with mutually developed standards and practices.
2. The Signatories intend to use their experience under the Accords to contribute to multilateral efforts to further develop international practices and rules applicable to preserving outer space heritage.

Section 10: Space Resources

1. The Signatories note that the utilization of space resources can benefit humankind by providing critical support for safe and sustainable operations.
2. The Signatories emphasize that the extraction and utilization of space resources, including any recovery from the surface or subsurface of the Moon, Mars, comets, or asteroids, should be executed in a manner that complies with the Outer Space Treaty and in support of safe and sustainable space activities. The Signatories affirm that the extraction of space resources does not inherently constitute national appropriation under Article II of the Outer Space Treaty, and that contracts and other legal instruments relating to space resources should be consistent with that Treaty.
3. The Signatories commit to informing the Secretary-General of the United Nations as well as the public and the international scientific community of their space resource extraction activities in accordance with the Outer Space Treaty.
4. The Signatories intend to use their experience under the Accords to contribute to multilateral efforts to further develop international practices and rules applicable to the extraction and utilization of space resources, including through ongoing efforts at the COPUOS.

Section 11: Deconfliction of Space Activities

1. The Signatories acknowledge and reaffirm their commitment to the Outer Space Treaty, including those provisions relating to due regard and harmful interference.

2. The Signatories affirm that the exploration and use of outer space should be conducted with due consideration to the United Nations Guidelines for the Long-term Sustainability of Outer Space Activities adopted by the COPUOS in 2019, with appropriate changes to reflect the nature of operations beyond low-Earth orbit.
3. Consistent with Article IX of the Outer Space Treaty, a Signatory authorizing an activity under these Accords commits to respect the principle of due regard. A Signatory to these Accords with reason to believe that it may suffer, or has suffered, harmful interference, may request consultations with a Signatory or any other Party to the Outer Space Treaty authorizing the activity.
4. The Signatories commit to seek to refrain from any intentional actions that may create harmful interference with each other's use of outer space in their activities under these Accords.
5. The Signatories commit to provide each other with necessary information regarding the location and nature of space-based activities under these Accords if a Signatory has reason to believe that the other Signatories' activities may result in harmful interference with or pose a safety hazard to its space-based activities.
6. The Signatories intend to use their experience under the Accords to contribute to multilateral efforts to further develop international practices, criteria, and rules applicable to the definition and determination of safety zones and harmful interference.
7. In order to implement their obligations under the Outer Space Treaty, the Signatories intend to provide notification of their activities and commit to coordinating with any relevant actor to avoid harmful interference. The area wherein this notification and coordination will be implemented to avoid harmful interference is referred to as a 'safety zone'. A safety zone should be the area in which nominal operations of a relevant activity or an anomalous event could reasonably cause harmful interference. The Signatories intend to observe the following principles related to safety zones:

(a) The size and scope of the safety zone, as well as the notice and coordination, should reflect the nature of the operations being conducted and the environment that such operations are conducted in;
(b) The size and scope of the safety zone should be determined in a reasonable manner

leveraging commonly accepted scientific and engineering principles;

(c) The nature and existence of safety zones is expected to change over time reflecting the status of the relevant operation. If the nature of an operation changes, the operating Signatory should alter the size and scope of the corresponding safety zone as appropriate. Safety zones will ultimately be temporary, ending when the relevant operation ceases; and
(d) The Signatories should promptly notify each other as well as the Secretary-General of the United Nations of the establishment, alteration, or end of any safety zone, consistent with Article XI of the Outer Space Treaty.

8. The Signatory maintaining a safety zone commits, upon request, to provide any Signatory with the basis for the area in accordance with the national rules and regulations applicable to each Signatory.
9. The Signatory establishing, maintaining, or ending a safety zone should do so in a manner that protects public and private personnel, equipment, and operations from harmful interference. The Signatories should, as appropriate, make relevant information regarding such safety zones, including the extent and general nature of operations taking place within them, available to the public as soon as practicable and feasible, while taking into account appropriate protections for proprietary and export-controlled information.
10. The Signatories commit to respect reasonable safety zones to avoid harmful interference with operations under these Accords, including by providing prior notification to and coordinating with each other before conducting operations in a safety zone established pursuant to these Accords.
11. The Signatories commit to use safety zones, which will be expected to change, evolve, or end based on the status of the specific activity, in a manner that encourages scientific discovery and technology demonstration, as well as the safe and efficient extraction and utilization of space resources in support of sustainable space exploration and other operations. The Signatories commit to respect the principle of free access to all areas of celestial bodies and all other provisions of the Outer Space Treaty in their use of safety zones. The Signatories further commit to adjust their usage of safety zones over time based on mutual experiences and consultations with each other and the international community.

Section 12: Orbital Debris

1. The Signatories commit to plan for the mitigation of orbital debris, including the safe, timely, and efficient passivation and disposal of spacecraft at the end of their missions, when appropriate, as part of their mission planning process. In the case of cooperative missions, such plans should explicitly include which Signatory has the primary responsibility for the end-of-mission planning and implementation.
2. The Signatories commit to limit, to the extent practicable, the generation of new, long-lived harmful debris released through normal operations, break-up in operational or post-mission phases, and accidents and conjunctions, by taking appropriate measures such as the selection of safe flight profiles and operational configurations as well as post-mission disposal of space structures.

Section 13: Final Provisions

1. Building on any consultative mechanisms in preexisting arrangements as appropriate, the Signatories commit to periodically consult to review the implementation of the principles in these Accords, and to exchange views on potential areas of future cooperation.

2. The Government of the United States of America will maintain the original text of these Accords and transmit to the Secretary-General of the United Nations a copy of these Accords, which is not eligible for registration under Article 102 of the Charter of the United Nations, with a view to its circulation to all the members of the Organization as an official document of the United Nations.
3. After October 13, 2020, any State seeking to become a Signatory to these Accords may submit its signature to the Government of the United States for addition to this text.

Adopted on October 13, 2020, in the English language.

Bibliography

Books – Papers – Monographs – Studies – Commentaries

ARNOULD Jacques, "Chapter 11: An Urgent Need to Explore Space", in: James S.J. Schwartz & Tony Milligan (Eds.), The Ethics of Space Exploration, Springer, 2016, p. 153 – 164.

AUGIER Mie & TEECE David J. (Eds.), The Palgrave Encyclopedia of Strategic Management, Palgrave MacMillan, 2018, 1846 p.

BARRAL Virginie, "Sustainable Development in International Law: Nature and Operation of an Evolutive Legal Norm", E.J.I.L, Vol. 23 no. 2, 2012, p. 377-400.

BARUAH Rishiraj & NANDINI Paliwal, "Sustainable Space Exploration and Use: Space Mining in Present and Future Perspectives", Proceedings of the International Institute of Space Law, 58, 2015, p. 23 - 42.

BASLAR Kemal, The Concept of the Common Heritage of Mankind in International Law, Nijhoff, 1998.

BENNOUNA Mohamed, "Les droits d'exploitation des ressources minérales des océans", *in:* BARDONET D. & VIRALLY M. (eds.), Le nouveau droit international de la mer, Paris, Pedone, p. 117-139.

BENSON James William, "Space Resources: First Come First Served", Proceedings on the Law of Outer Space, Vol. 41, 1998, p. 46-49.

BERGAMASCO Federico, "Space Mining and the Protection of Extra-Terrestrial Environment in the Light of Article IX of the Outer Space Treaty", Proceedings of the International Institute of Space Law, 60, 2017, pp. 157-170.

BERNÁRDEZ Santiago Torres, "(Chapter) 37 - Interpretation of Treaties by the International Court Of Justice Following the Adoption of the 1969 Vienna Convention on the Law of Treaties", Liber Amicorum Professor Ignaz Seidl-Hohenveldern, Brill, 1998, p. 721-748.

BERNER Katharina, "Authentic Interpretation in Public International Law", ZaöRV, Vol. 76, 2016, p. 845-878.

BITTENCOURT NETO Olavo de O., HOFMANN Mahulena, T. MASSON-ZWAAN & D. STEFOUDI (eds.), Building Blocks for the Development of an International Framework for the Governance of Space Resource Activities: A Commentary, Eleven International Publishing, 2020, 164 p.

G. D. Kyriakopoulos, *Space Resource Utilization and Exploitation: The Suspended Step of International Law*, Space and Society,
https://doi.org/10.1007/978-3-031-96336-0

BOHACEK Petr, WORDEN Simon P. & GRATTAN Kyran, "Benefit-Sharing as Investment Protection for Space Resource Utilization", New Space, Volume 10, Number 2, 2022, https://doi.org/10.1089/space.2021.0050.

BOYLE Alan E., "Dispute Settlement and the Law of the Sea Convention: Problems of Fragmentation and Jurisdiction", ICLQ, vol. 46, 1997, p. 37-54.

BRESARD Olivier, "Les aspects financiers de l'exploitation des ressources de fonds marins internationaux", *in:* SFDI, Perspectives du droit de la mer à l'issue de la 3e Conférence des Nations unies, Colloque de Rouen, Pedone, Paris, 1984.

BRITTINGHAM Bryon C., "Does the World Really Need New Space Law?", Oregon R.I.L., vol. 12, 2010.

BRUHNS Sara & HAQQ-MISRA Jacob, "A pragmatic approach to sovereignty on Mars", Space Policy, Vol. xxx, 2016, p. 1-7.

BUECKLING Adrian, "The Strategy of Semantics and the 'Mankind Provisions' of the Space Treaty", J.S.L., vol. 7 no 1, 1979.

CAPURSO Andrea, "The Non-Appropriation Principle: A Roman Interpretation", Proceedings of the International Institute of Space Law, Vol. 61, 2018, p. 111-128.

CHATZIPANAGIOTIS Michael, "3D Printing Using Material from Celestial Bodies: A Method to Circumvent the Non-Appropriation Principle", Proceedings of the International Institute of Space Law, Vol. 59, 2016, p. 247-258.

CHENG Bin, Studies in International Space Law, Clarendon Press, Oxford, 1997.

CHRISTOL Carl Q. "Article 2 of the 1967 Principles Treaty Revisited" A.A.S.L., Vol. 9, 1984.

CHURCHILL Robin Rolf & LOWE Alan Vaughan, The Law of the Sea, 3rd Edition, 1999, Manchester University Press: Juris Publishing, 494 p.

CINELLI Claudia & POGORZELSKA Katarzyna, The Current International Legal Setting for the Protection of the Outer Space Environment: the Precautionary Principle Avant La Lettre, Reciel 22 (2), 2013.

CRAWFORD James, Brownlie's Principles of Public International Law, Oxford University Press, 9th Edition, 2019, 872 p.

DAVIS Chelsey & SUNDAHL Mark J. "The Hague Working Group on Space Resources: Creating the Legal Building Blocks for a New Industry", The Air & Space Lawyer, Vol. 30, No 3, Fall 2017, 5p.

DE MAN Philip, "Interpreting the UN Space Treaties as the Basis for a Sustainable Regime of Space Resource Exploitation", *in :* Kyriakopoulos G.D. & Manoli M. (eds.), The Space Treaties at Crossroads: Considerations de Lege Ferenda, Springer, 2019, p. 15-33.

DE MAN Philip, Exclusive Use in an Inclusive Environment - The Meaning of the Non-Appropriation Principle for Space Resource Exploitation, Springer, 2016, 481 p.

DE MAN Philip, "Commercial Exploitation of Outer Space and Celestial Bodies - A Functional Solution to the Natural Resource Challenge", Proceedings of the International Institute of Space Law, Vol. 53, 2010, p. 28-38.

DEPLANO Rossana, "Inclusive Space Law: The Concept of Benefit Sharing in the Outer Space Treaty", I.C.L.Q., Vol. 72, July 2023, p. 671–714.

DEPLANO Rossana, "The Artemis Accords: Evolution or Revolution in International Space Law?", I.C.L.Q., Vol. 70, no. 3, July 2021, p. 799-820.

DÖRR Oliver & SCHMALENBACH Kirsten (eds.), Vienna Convention on the Law of Treaties - A Commentary, Springer, 2nd Edition, 2018, 1535 p.

DUPUY Pierre-Marie & KERBRAT Yann, Droit International Public, 13e ed., 2016.

DUPUY René-Jean, L'océan partagé, Pedone, Paris, 1979, 287 p.

FREELAND Steven & JAKHU Ram, "Article II", *in* S. Hobe, B. Schmidt-Tedd, K.-U. Schrogl & G. Meishan Goh (Eds.), Cologne Commentary on Space Law, Vol. 1 Outer Space Treaty, Carl Heymanns Verlag, 2009.

GABRYNOWICZ Joanne Irene, "The 'province' and 'heritage' of mankind reconsidered: A new beginning", NASA Johnson Space Center, 2nd Conference on Lunar Bases and Space Activities of the 21st century, vol. 2, 1992.

GOEDHUIS Daniel, "Some Recent Trends in the Interpretation and Implementation of the Rules of International Space Law", Columbia J.T.L., vol. 19, 1981.

GOROVE Stephen, "The Concept of the Common Heritage of Mankind: A Political, Moral, and Legal Innovation?", San Diego L.R., vol. 9, 1972.

HOBE Stephan, "The Artemis Accords: What They Mean for the Development of International Space Law", Zeitschrift fur Luft- und Weltraumrecht (German Journal of Air and Space Law), vol. 70, no. 1, 2021, p. 1-12.

HOBE Stephan, "The Impact of New Developments on International Space Law (New Actors, Commercialization, Privatization, Increase in the Number of "Space-faring Nations"), Uniform Law Review, 2010, Vol. 15, issue 3-4.

HOBE Stephan, "Article I", *in:* Stephan Hobe, Bernhard Schmidt-Tedd & Kai-Uwe Schrogl (Eds.), CoCOSL – Cologne Commentary on Space Law, Volume 1, Outer Space Treaty, Carl Heymanns Verlag, 2009, p. 34.

HOBE Stephan, "Adequacy of the Current Legal and Regulatory Framework Relating to the Extraction and Appropriation of Natural Resources", *in :* Policy and Law Relating to Outer Space Resources: Examples of the Moon, Mars, and other Celestial Bodies, Institute of Air and Space Law, McGill University Workshop Proceedings, 28-30 June 2006.

HOBE Stephan, "The Legal Framework for a Lunar Base Lex Data and Lex Ferenda", *in:* G. Lafferranderie & D. Crowther (eds.), Outlook on Space Law over the Next 30 Years: Essays Published for the 30th Anniversary of the Space Treaty, Springer, 1997.

HOFMANN Mahulena, "COSPAR Recommendations in a New Context: Environmental Aspects of Space Mining", Proceedings of the International Institute of Space Law, 59, 2016, pp. 201-212.

HYTREK Matthew, "Property Rights in Current Space Law: A Hindrance to Space Exploration", Whittier Law Review, vol. 39, no. 1, Spring 2018, p. 90-114.

JAECKEL Aline, "Benefitting from the Common Heritage of Humankind: From Expectation to Reality", The International Journal of Marine and Coastal Law, Vol. 35, 2020, p. 1-22.

JAKHU Ram, "Twenty Years of the Moon Agreement: Space Law Challenges for Returning to the Moon", Z.L.W., vol. 54, 2005a.

JAKHU Ram, "United Nations Principles in Outer Space", *in:* Proceedings of the United Nations/Nigeria Workshop on Space Law on "Meeting international responsibilities and addressing domestic needs, held on 21-24 November 2005b, in Abuja, Nigeria.

JOHNSON Christopher D., "Renewed Ambitions in Space-Exploration Lawmaking", Air and Space Lawyer, vol. 33, no. 3, 2020, p. 19-21.

KEEFE Heidi, "Making the Final Frontier Feasible: A Critical Look at the Current Body of Outer Space Law", Santa Clara High Technology Law Journal, Vol. 11 Issue 2, 1995.

KLABBERS Jan, The Concept of Treaty in International Law, Kluwer Law International, The Hague/London/Boston, 1996, 307 p.

KYRIAKOPOULOS George D., "Colonies on the Moon (and/or Mars)? New Challenges for International and National Law", Proceedings of the International Institute of Space Law, 2020, p. 169-181.

KYRIAKOPOULOS George D., "Positive Space Law and Privatization of Outer Space: Fundamental Antinomies", *in:* Kyriakopoulos G.D. & Manoli M. (eds.), The Space Treaties at Crossroads: Considerations de Lege Ferenda, Springer, 2019, p. 1-13.

KYRIAKOPOULOS George D., "Legal Challenges Posed by the Action of Non-State Actors in Outer Space", *in:* Manoli M. & Habchi S.B. (eds.), Monograph Series V – Conflicts in Space and the Rule of Law, McGill Centre for Research in Air and Space Law, June 2017.

KYRIAKOPOULOS George D., "Where Law Meets Cinema: James Cameron's Avatar as Food for Thought About the Anthropocentric Nature of Space Law", Proceedings of the international Institute of Space Law, 58, 2015, pp. 303-318.

KYRIAKOPOULOS Georges, "Statut des espaces maritimes et régime juridique de la navigation aérienne", RGDIP, 2001/2, p. 307-329.

LACHS Manfred, "Pacta Sunt Servanda", *in*: 7 ENCYCLOPEDIA OF PUBLIC INTERNATIONAL LAW, 1981.

LARSEN Paul B., "Is There a Legal Path to Commercial Mining on the Moon?", University of Pittsburgh Law Review, vol. 83, no. 1, Fall 2021, p. 1-50.

LEE Ricky J., "Saudi Arabia's Withdrawal from the Moon Agreement", Australian and New Zealand Society of International Law, 25 July 2023, https://anzsilperspective.com/saudi-arabias-withdrawal-from-the-moon-agreement/.

LEE Ricky J., Law and Regulation of Commercial Mining of Minerals in Outer Space, Springer, 2012.

LETERRE Gabriel, "Space Mining and Environmental Protection: Recycling International Agreements into New Legal Practices", Proceedings of the International Institute of Space Law, 62, 2019, p. 83-94.

LÉVY Jean-Pierre, La Conférence des Nations Unies sur le droit de la mer (Histoire d'une négociation singulière), Pedone, Paris, 1983, 159 p.

LISTNER Michael J., "Four key points regarding Saudi Arabia's withdrawal from the Moon Agreement", The Space Review, Monday, December 11, 2023, https://www.thespacereview.com/article/4706/1/.

LIU Hao & TRONCHETTI Fabio, "The Exclusive Utilization Space: A New Approach to the Management and Utilization of the Near Space", U. Pa. J. Int'l L., Vol. 40:3, 2019, p. 537-574.

LOTHIAN Sarah, "How does the new BBNJ Agreement regulate access to and benefit-sharing of marine genetic resources at deep-sea hydrothermal vents?", 1st June 2023, ECO LAW Blog, https://blogs.ed.ac.uk/ecolaw/2023/06/01/how-does-the-new-bbnj-agreement-regulate-access-to-and-benefit-sharing-of-marine-genetic-resources-at-deep-sea-hydrothermal-vents/ (Last accessed: 26.1.2025).

MALAGAR Leo B. & MAGDOSA-MALAGAR Marlo Apalisok, "International Law of Outer Space and the Protection of Intellectual Property Rights", Boston U.I.L.J., vol. 17, 1999.

MANOLI Maria, "Pondering the Legitimacy of the Outer Space Treaty from the Perspective of Space Natural Resources Exploitation: Is it Time to Act instead of React?", Proceedings of the International Institute of Space Law, Vol. 60, 2017, p. 267-280.

MANOLI Maria, The Immaterial Territoriality of Outer Space: Reconstructing Territorylessness in Modern Space Governance (Montreal: McGill University, 2021), online: EScholarship https://escholarship.mcgill.ca/concern/theses/hd76s530m, 323 p.

MARBOE Irmgard, "Reviewing the Moon Agreement or Amending the Outer Space Treaty? - Views of UNCOPUOS Member States", Proceedings of the International Institute of Space Law, Vol. 62, 2019, p. 399-416.

MARBOE Irmgard, "The End of the Concept of Common Heritage of Mankind", Proceedings of the International Institute of Space Law, Vol. 59, 2016, p. 225-238.

MARCHISIO Sergio, "Article IX", *in:* Hobe S., Schmidt-Tedd B., Schrogl K.-U., Goh M. (eds.), Cologne Commentary on Space Law, Vol. 1, Carl Heymanns Verlag, 2009.

MATUSIAK Michal, "Artemis Accords – A new era of space law or an initiative of the World's superpower?", Studia Juridica, XCVII, 2023, p. 168-184.

MAYR Teresa F. & MAYR-SINGER Jelka, "Keep the Wheels Spinning: The Contributions of Advisory Opinions of the International Court of Justice to the Development of International Law", ZaöRV 76, 2016.

MEYER Z., "Private Commercialization of Space in an International Regime: A Proposal for a Space District", Northwestern J.I.L.&B., Vol. 30 Issue 1, 2010.

MINEIRO Michael, Space Technology Export Controls and International Cooperation in Outer Space, Springer, 2012.

MOSTESHAR Sa'id, "Artemis: The Discordant Accords", Journal of Space Law, vol. 44, No 2, 2020, p. 591-603.

NEEF Rachel, "Artemis Accords: A New Path Forward for Space Lawmaking?", Adelaide Law Review, vol. 42, no. 2, 2021, p. 569-580.

OSADA Yutaka, "Governance of Space Resources Activities: In the Wake of the Artemis Accords", Georgetown Journal of International Law, vol. 53, no. 3, Spring 2022, p. 399-511.

PAZARTZIS Photeini, "Judicial Activism and Judicial Self-Restraint: The PCIJ's Lotus Case" *in:* Tams ChJ., Fitzmaurice. & Merkouris P. (eds.), Legacies of the Permanent Court of International Justice, Nijhoff, 2013.

PELLET A., "Lotus que des sottises on profère en ton nom! Remarques sue le concept de souveraineté dans la jurisprudence de la Cour mondiale", *in:* Mélanges en l'honneur de Jean-Pierre Puissochet, L'Etat souverain dans le monde d'aujourd'hui, Pedone, 2008.

POP Virgiliu, Who Owns the Moon? Extraterrestrial Aspects of Land and Mineral Resources Ownership, Springer, 2009, 175 p.

PORRAS Daniel A., "The 'Common Heritage' of Outer Space: Equal Benefits for most of Mankind", California West.I.L.J., vol. 37, no 1, 2006.

RANGANATHAN Surabhi, "Global Commons", European Journal of International Law, Volume 27, Issue 3, 1 August 2016.

REINHOLD Steven, "Good Faith in International Law", UCLJLJ vol. 2, 2013.

REUTER Paul, Introduction au droit des traités, PUF, 3e éd., 1995.

RIBBELINK Olivier, "Article III", *in:* Hobe S., Schmidt-Tedd B., Schrogl K.-U., Goh M. (eds.), Cologne Commentary on Space Law, Vol. 1, Outer Space Treaty, Carl Heymanns Verlag, 2009.

ROSENNE Shabtai, "The perplexities of modern international law : general course on public international law", R.C.A.D.I., vol. 291, 2001, p. 9-471.

ROUCOUNAS Emmanouil, Dimossio Diethnes Dikaio (Public international law) (in Greek), Nomiki Bibliothiki, 3rd Edition, 2091, and 4th edition, 2021.

ROUSSEAU Charles, Droit international public, Tome I – Introduction et sources, Sirey, Paris, 1970, 464 p.

SCHACHTER Oscar, "The Twilight Existence of Nonbinding International Agreements", AJIL, Vol. 71, No 2 (Apr. 1977), pp. 296-304.

SCHEMBRI Rebecca, "The Outer Space Treaty Needs Amending", Human Space Program, January 25, 2023, https://medium.com/@humanspaceprogram/the-outer-space-treaty-needs-amending-d9813c40ba1b

SCHRIJVER Nico & WEISS Friedl (eds.), International Law and Sustainable Development: Principles and Practice, Nijhoff, Leiden-Boston, 2004, 711 p.

SCHROGL Kai-Uwe, "We Must Not Overrate the Artemis Accords", in: SpaceWatch Global, https://spacewatch.global/2020/11/spacewatchgl-interviews-kai-uwe-schrogl-we-must-not-overratethe-artemis-accords/.

SCOVAZZI Tullio, "The evolution of international law of the sea: new issues, new challenges", R.C.A.D.I., Vol. 286, 2000, p. 39-243.

SMITH Walker A., "Using the Artemis Accords to Build Customary International Law: A Vision for a U.S.-Centric Good Governance Regime in Outer Space", Journal of Air Law and Commerce, vol. 86, no. 4, 2021, p. 661-700.

STERNS Patricia M. & TENNEN Leslie I., "Privatering and Profitering on the Moon and Other Celestial Bodies: Debunking the Myth of Property Rights in Space", Adv. Space Res., Vol. 31, No 11, 2003.

TANAKA Yoshifumi, The International Law of the Sea, 3rd ed., Cambridge UP, 2019, p. 149.

TASHJIAN Joel S., "Contentious Matters and the Advisory Power: The ICJ and Israel's Wall", Chicago J.I.L., Vol. 6 No 1, 6.1.2005, Article 25.

TENNEN Leslie, "Enterprise Rights in Extraterrestrial Resources: Commentary on Outer Space and International Geography: Article II and the Shape of Global Order", New England Law Review, Vol. 52, No. 2, Winter 2018, pp. 139-158.

TENNEN Leslie I., "Enterprise Rights and the Legal Regime for Exploitation of Outer Space Resources", U. Pac. L. Rev., Vol. 47, 2017, p. 281-299. Available at: https://scholarlycommons.pacific.edu/uoplawreview/vol47/iss2/14.

TENNEN Leslie I., "Towards a New Regime for Exploitation of Outer Space Mineral Resources", Nebraska Law Review, Vol. 88, No. 4, 2010, p. 794-831.

TENNEN Leslie I., "New Developments and the Legal Framework Covering the Exploitation of the Resources of the Moon", Proceedings on the Law of Outer Space, Vol. 47, 2004a, p. 520-529.

TENNEN Leslie I., "Second Commentary on Emerging System of Property Rights in Outer Space", Proceedings of the United Nations/Republic of Korea Workshop on Space Law "United Nations treaties on outer space: actions at the national level", Daejeon, Republic of Korea, 3-6 November 2003, 2004b, ST/SPACE/22, p. 342-347.

TRONCHETTI Fabio & LIU Hao, "Australia's Signing of the Artemis Accords: A Positive Development or a Controversial Choice?", AUSTL. J. INT'L AFF., Vol. 75, 2021, p. 243-251.

TRONCHETTI Fabio, The Exploitation of Natural Resources of the Moon and Other Celestial Bodies - A Proposal for a Legal Regime, Martinus NIjhoff, Leiden – Boston, 2009.

VERESHCHETIN V.S. & DANILENKO G.M., "Custom as a source of International Law of Outer Space", J. Space L., Vol. 13, No 1, 1985.

VILLIGER Mark E., Commentary on the 1969 Vienna Convention on the Law of Treaties, Brill/Nijhoff, Leiden, The Netherlands, 2008.

VOIGT Christina, Sustainable Development as a Principle of International Law: Resolving Conflicts between Climate Measures and WTO Law, NIjhoff, Leiden-Boston, 2009, 426 p.

VOLYNSKAYA Olga, "Space Resources Exploitation from the international and Domestic Law Perspectives: The Russian Approach", Proceedings of the International Institute of Space Law, Vol. 59, 2016, p. 291-300.

VON DER DUNK Frans (ed.) (with F. Tronchetti), Handbook of Space Law, Edward Elgar, 2015.

VON DER DUNK Frans G., "The Artemis Accords as a Tool of Cooperation", Proceedings of the International Institute of Space Law, 2021, p. 145-157.

VONNEGUT Kurt, The Sirens of Titan, 1959.

WANG Guoyu, "NASA's Artemis Accords: the path to a united space law or a divided one?", The Space Review, Monday, August 24, 2020, https://www.thespacereview.com/article/4009/1/.

WILLIS Jared E., "Sovereignty in Space: Finding a Source of Private Property Rights in the Final Frontier", The Elon Law Journal, Vol. 14, 2022, p. 361-385.

WHITE Jr. Wayne N., "Real Property Rights in Outer Space The Future Applications of the Outer Space Treaty", Proceedings on the Law of Outer Space, Vol. 40, 1997, p. 370-384.

WILDE Daniel, LILY Hannah, CRAIK Neil & CHAKRABORTY Anindita, "Equitable sharing of deep-sea mining benefits: More questions than answers", Marine Policy, Vol. 151, 2023.

XU Fengna & SU Jinyuan, "New Elements in The Hague Space Resources Governance Working Group's Building Blocks", Space Policy, Vol. 53, article id. 101386, August 2020, 9 p.

XU Fengna & SU Jinyuan, "Towards a legal regime of benefits sharing for space mining: With some experience from the Area", Resources Policy, Volume 76, 2022, article id. 102627, 10.1016/j.resourpol.2022.102627.

XU Fengna, SU Jinyuan & MEHDI Miqdad, "A Re-Examination of Fundamental Principles of International Space Law at the Dawn of Space Mining", Journal of Space Law, Vol. 44, No 1, 2020, p. 1-43.

YOKARIS Angelos & KYRIAKOPOULOS George, International Law of Airspace, Outer Space, Nomiki Vivliothiki, 2020, (in Greek).

YOKARIS Angelos, International Air and Space Law (in Greek), Ant. N. Sakkoulas, Athens, 1996, 314 p.

ZELL Jeremy L., "Putting a Mine on the Moon: Creating an International Authority to Regulate Mining Rights in Outer Space", Minnesota J.I.L., vol. 15, 2006.

ZULLO Kelly M., "The Need to Clarify the Status of Property Rights in International Space Law", Georgetown L.J., vol. 90, 2001-2002.

Documents

International Conventions

2023 Agreement under the United Nations Convention on the Law of the Sea on the Conservation and Sustainable Use of Marine Biological Diversity of Areas beyond National Jurisdiction, New York, 19 June 2023, open for signature on 20 September 2023 - until 20 September 2025. Not yet in force.

2020 The Artemis Accords: Principles for Cooperation in the Civil Exploration and Use of the Moon, Mars, Comets, and Asteroids for Peaceful Purposes', National Aeronautics and Space Administration (13 October 2020) ('Artemis Accords').

2010 Nagoya Protocol on Access to Genetic Resources and the Fair and Equitable Sharing of Benefits Arising from their Utilization to the Convention on Biological Diversity, adopted on 29 October 2010, entered into force on 12 October 2014, 3008/U.N.T.S./3.

1994 Agreement relating to the Implementation of Part XI of the United Nations Convention on the Law of the Sea of 10 December 1982, adopted on 28/07/1994, entered into force on 28/ Convention 07/1996, 1836 UNTS 3, 33 ILM 1309 (1994), UN Doc. A/RES/48/263.

1992 Convention on Biological Diversity, adopted on 5 June 1992 (Rio de Janeiro), entered into force on 29 December 1993, 1760/U.N.T.S./79, 31 I.L.M. 818 (1992).

1982 Convention on the Law of the Sea, adopted on 10 December 1982, entered into force on 16 November 1994. 1833/U.N.T.S./397

1979 Agreement Governing the Activities of States on the Moon and Other Celestial Bodies, adopted on 5 December 1979, opened for signature on 18 December 1979, entered into force on 11 July 1984. 1363/U.N.T.S./3.

1970 Declaration of Montevideo on the Law of the Sea, signed at Montevideo, Uruguay, May 8, 1970, International Legal Materials, Volume 9, Issue 5, September 1970, pp. 1081 – 1083.

1969Vienna Convention on the Law of Treaties, done at Vienna on 23 May 1969, entered into force Jan. 27, 1980, 1155 U.N.T.S. 331.

1967 Treaty on Principles Governing the Activities of States in the Exploration and Use of Outer Space, including the Moon and Other Celestial Bodies, entered into force Oct. 10, 1967, 18 U.S.T. 2410, 610 U.N.T.S. 205.

1946 Convention on the Privileges and Immunities of the United Nations, adopted by the General Assembly of the United Nations on 13 February 1946, entered into force in 17 September 1946, U.N.T.S. 1, 15, and 90, 327 (corrigendum to vol. 1).

1946 United Nations, Statute of the International Court of Justice, 18 April 1946.

1945 United Nations, Charter of the United Nations, 24 October 1945, 1 UNTS XVI.

1944 Convention on International Civil Aviation, adopted on 7 December 1944, entered into force on 4 April 1947, 15/U.N.T.S./295.

National Laws

(Japan) Government of Japan, Act on the Promotion of Business Activities Related to the Exploration and Development of Space Resources, The Space Resources Act, Law No. 83 of 2021, art. 2, 3 (June 15th 2021).

(United Arab Emirates) Government of the United Arab Emirates, Federal Law No. (12) of 2019 Issued on 19/12/2019 Corresponding to 22 Rabi' Al-Akhar 1441H. ON THE REGULATION OF THE SPACE SECTOR, art. 4 (December 19th 2019).

(United Arab Emirates) Government of the United Arab Emirates, National Space policy of the United Arab Emirates, §§ 5.1.6, 2.2., 5.1.6., 5.2.4., 5.2.5., 5.3.5 (2019).

(USA) DCPD-202000245 - Executive Order 13914-Encouraging International Support for the Recovery and Use of Space Resources, April 6, 2020.
(Greece) Law 4508/2017 (Government Gazette 200/A′/22-12-2017) on “Authorization of space activities -Registration in the National Register of Space Objects- Establishment of a Greek Space Organization and other provisions”, as amended by Law 4712/2020/146/A′/29-07-2020.
(Luxembourg) Loi du 20 juillet 2017 sur l’exploration et l’utilisation des ressources de l’espace, Signature : 20.07.2017, Publication : 28.07.2017, Entrée en vigueur : 01.08.2017.
(USA) H.R.2262 - U.S. Commercial Space Launch Competitiveness Act 114th Congress (2015-2016).
(USA) H.R.3942 - Commercial Space Launch Act 98th Congress (1983-1984).

United Nations Documents

Secretary-General

Secretary-General Policy Brief on Outer Space: Our Common Agenda Policy Brief 7 For All Humanity – the Future of Outer Space Governance (May 2023), A/77/CRP.1/Add.6, https://www.unoosa.org/oosa/en/documents/doc/policy-brief.html (last accessed: 13.2.2024).

General Assembly

A/RES/77/121, “International cooperation in the peaceful uses of outer space”, 12 December 2022.
A/RES/76/3, “The ‘Space2030’ Agenda: space as a driver of sustainable development”, 25 October 2021.
A/RES/74/82, “International cooperation in the peaceful uses of outer space”, 26 December 2019.
A/RES/62/217, “International cooperation in the peaceful uses of outer space”, 22 December 2007.
A/RES/51/122, “Declaration on International Cooperation in the Exploration and Use of Outer Space for the Benefit and in the Interest of All States, Taking into Particular Account the Needs of Developing Countries”, 13 December 1996.
A/RES/47/68, “Principles Relevant to the Use of Nuclear Power Sources in Outer Space”, 14 December 1992.
A/RES/41/65, “Principles relating to remote sensing of the Earth from outer space”, 3 December 1986.
A/RES/37/92, “Principles Governing the Use by States of Artificial Earth Satellites for International Direct Television Broadcasting”, 10 December 1982.
Statement of Arvid Pardo, 1 November 1967, First Committee, UNGA, 22 UN GAOR, 1515th & 1516th Meeting, p. 8-9 (1515th) and 2 (1516th) (emphasis by the author).
RES 1348 XIII, “Question of the Peaceful Use of Outer Space”, 13 December 1958.

Security Council

S/RES/781 (1992), 9 October 1992.
S/RES/748 (1992), 31 March 1992.
S/RES/688 (1991), 5 April 1991.
S/RES/670 (1990), 25 September 1990.

United Nations Committee on the Peaceful Uses of Outer Space

A/AC.105/C.2/2022/CRP.21, 1.4.2022.
A/AC.105/C.2/2022/CRP.13, 28.3.2022.
A/AC.105/L.326, 25.6.2021.
UNOOSA, Guidelines for the Long-term Sustainability of Outer Space Activities of the Committee on the Peaceful Uses of Outer Space, United Nations, Vienna, 2021.
A/AC.105/C.2/L.315, 3.2.2020.
A/AC.105/C.2/L.311, 4.3.2019.
A/AC.105/2019/CRP.17, 19.6.2019.
A/AC.105/C.2/2018/CRP.18, 12.4.2018.
A/AC.105/C.2/L.299, 27.1.2017.
Report of the Legal Subcommittee on its 55th session, held in Vienna from 4 to 15 April 2016, UN Doc. A/AC.105/1113.
UNOOSA, Space Debris Mitigation Guidelines of the Committee on the Peaceful Uses of Outer Space, United Nations, Vienna, 2010.

Miscellaneous

Building Blocks for the Development of an International Framework on Space Resource Activities, The Hague International Space Resources Governance Working Group, November 2019, available at https://www.universiteitleiden.nl/en/law/institute-of-public-law/institute-of-air-space-law/the-hague-space-resources-governance-working-group (last accessed: 28.04.2023).

Jurisprudence

International Court of Justice

ICJ, Maritime Dispute (Peru v. Chile) (Judgment), I.C.J. Reports 2014, p. 3.
ICJ, Accordance with International Law of the Unilateral Declaration of Independence in Respect of Kosovo, Advisory Opinion, I.C.J. Reports 2010, p. 403.
ICJ, Legal Consequences of the Construction of a Wall in the Occupied Palestinian Territory, Advisory Opinion, I.C.J. Reports 2004, p. 136.
ICJ, Sovereignty over Pulau Ligitan and Pulau Sipadan (Judgment), I.C.J. Reports 2002, p. 625.
ICJ, Legality of the Threat or Use of Nuclear Weapons, Advisory Opinion, 1.C.J. Reports 1996, p. 226.
ICJ, Kasikili/Sedudu Island (Botswana v. Namibia) , Judgment, I.C.J. Reports 1999, p. 1045.
ICJ, Military and Paramilitary Activities in and against Nicaragua (Nicaragua v. United States of America). Merits, Judgment. I.C.J. Reports 1986, p. 14.
ICJ, Aegean Sea Continental Shelf, Judgment, I.C.J. Reports 1978, p. 3.
ICJ, Legal Consequences for States of the Continued Presence of South Africa in Namibia (South West Africa) notwithstanding Security Council Resolution 276 (1970), Advisory Opinion, I.C.J. Reports 1971, p. 16.
ICJ, North Sea Continental Shelf, Judgment, I.C.J. Reports 1969, p. 3.
ICJ, South West Africa Cases (Ethiopia v. South Africa; Liberia v. South Africa), Preliminary Objections, Judgment of 21 December 1962: I.C.J. Reports 1962, p. 319.
ICJ, Certain Expenses of the United Nations (Article 17, paragraph 2, of the Charter), Advisory Opinion, I.C.J. Reports 1961.

ICJ, Conditions of Admission of a State to Membership in the United Nations, Advisory Opinion, I.C.J. Reports 1948.

Permanent Court of International Justice

PCIJ, S.S. Lotus (Fr. v. Turk.), 1927 P.C.I.J. (ser. A) No. 10 (Sept. 7), Publications of the Permanent Court of International Justice, Series A - No. 10; Collection of Judgments, A.W. Sijthoff, Leyden, 1927.

International Tribunal for the Law of the Sea

ITLOS, Responsibilities and obligations of States with respect to activities in the Area, Advisory Opinion, 1 February 2011, ITLOS Reports 2011, p. 10.

National Courts

US DISTRICT COURT, DISTRICT OF NEVADA, Nemitz v. United States of America et al., Dismissal of Plaintiff's claim, with prejudice, of 26 April 2004.

Index

G. D. Kyriakopoulos, *Space Resource Utilization and Exploitation: The Suspended Step of International Law*, Space and Society, https://doi.org/10.1007/978-3-031-96336-0

GPSR Compliance

The European Union's (EU) General Product Safety Regulation (GPSR) is a set of rules that requires consumer products to be safe and our obligations to ensure this.

If you have any concerns about our products, you can contact us on ProductSafety@springernature.com

In case Publisher is established outside the EU, the EU authorized representative is:

Springer Nature Customer Service Center GmbH
Europaplatz 3
69115 Heidelberg, Germany

Batch number: 09212143

Printed by Printforce, the Netherlands